# SpringerBriefs in Materials

The SpringerBriefs Series in Materials presents highly relevant, concise monographs on a wide range of topics covering fundamental advances and new applications in the field. Areas of interest include topical information on innovative, structural and functional materials and composites as well as fundamental principles, physical properties, materials theory and design.

**Indexed in Scopus (2022).**

SpringerBriefs present succinct summaries of cutting-edge research and practical applications across a wide spectrum of fields. Featuring compact volumes of 50 to 125 pages, the series covers a range of content from professional to academic. Typical topics might include

- A timely report of state-of-the art analytical techniques
- A bridge between new research results, as published in journal articles, and a contextual literature review
- A snapshot of a hot or emerging topic
- An in-depth case study or clinical example
- A presentation of core concepts that students must understand in order to make independent contributions

Briefs are characterized by fast, global electronic dissemination, standard publishing contracts, standardized manuscript preparation and formatting guidelines, and expedited production schedules.

Rachid Masrour

# Magnetoelectronic, Optical, and Thermoelectric Properties of Perovskite Materials

 Springer

Rachid Masrour
Laboratory of Solid Physics
Faculty of Sciences Dhar El Mahraz
Sidi Mohamed Ben Abdellah University
Fez, Morocco

ISSN 2192-1091            ISSN 2192-1105   (electronic)
SpringerBriefs in Materials
ISBN 978-3-031-48966-2       ISBN 978-3-031-48967-9   (eBook)
https://doi.org/10.1007/978-3-031-48967-9

This Springer imprint is published by the registered company Springer Nature Switzerland AG
The registered company address is: Gewerbestrasse 11, 6330 Cham, Switzerland

Paper in this product is recyclable.

# General Introduction

The perovskite oxide is a versatile class of compounds with A-site being 12 coordinated and occupied by rare earth or alkaline earth ions with large ionic radii. The B-site being octahedrally coordinated and generally occupied by a transition metal ion or rare earth ion with smaller ionic radii. It contains corner sharing $BO_6$ octahedra and the voids are filled by the larger radius A cations. Perovskite solar cells have attracted extensive attention due to their excellent optoelectronic properties, low cost, and easy fabrication, among other qualities [1–3]. Because of lead's high toxicity for the environment and humans, the industrial application of lead-based perovskites has been greatly limited [4, 5]. Currently, perovskite solar cells hold a record efficiency of 25.6% in a laboratory scale [6]. For organic-inorganic hybrid perovskites, secondary growth is usually achieved by a post-synthetic treatment induced in-situ transformation to perovskites with reduced dimension and higher stability, especially along the defect-rich grain boundaries [7]. Inorganic perovskites, exemplified by $CsPbI_3$, are characterized by superior chemical stability compared to their organic A-site cation counterparts. However, they do exhibit a higher density of defects. Nevertheless, achieving an organized secondary arrangement or reconstruction in these inorganic perovskites with reduced defects presents a significant challenge due to their distinctive chemical properties [8]. The performance of perovskite photodetectors has made a breakthrough in a short time. At present, the perovskite photodetectors with external quantum efficiency over 90% have been realized [8], and the external quantum efficiency (EQE) can be further enhanced by introducing photomultiplication (PM) effect. Moreover, some special properties, such as oxygen vacancies and the change in the B-site ion valence state, will significantly influence the catalytic activity of these perovskite materials [9]. Additionally, perovskite oxides can be made into different types of sensors, e.g., current and capacitance sensors, which are based on their various electrochemical properties such as good conductivity and dielectric properties [10, 11]. The research in the area of A-site ordering of this family of double perovskites is less compared to the B-site ordering. The result of these rare studies of A-site ordering is particularly due to the fact that most of the materials with these sorts of ordering needs high pressure synthesis and exist over a very narrow range of temperature [12–14]. Previous studies have shown that

$La_2NiTiO_6$ and $La_2MnTiO_6$ exhibit long-range paramagnetism at room temperature, which can explain laterally that Ti does not exhibit antiferromagnetism macroscopically in the double perovskite system, eliminating the risk of introducing additional antiferromagnetic particles [15, 16]. Previous researchers have found many interesting properties in halide perovskites [17–19], due to their complicated structure and noncentral symmetry, especially when transition metal (TM) ions are incorporated or used [20–22]. Recently, chalcogenide perovskites have received increased attention as an emerging ionic semiconductor family with fascinating optoelectronic properties [23–28]. The electronic, magnetic and magnetocaloric properties of $T_{1-x}Sr_xMnO_3$ with (T=La, Pr) and (x = 0.35 and 0.25) were studied by using the density functional theory and Monte Carlo study with the generalized gradient GGA, GGA+U (Hubbard parameter) and the Modified Becke-Johnson approximation. The Monte Carlo simulations were used to calculate some magnetic parameters such as the exchange coupling constants and magnetic moments. The electronic, optical, thermoelectric, magnetic, and magnetic properties of the $GdCrO_3$ system by using the DFT and Monte Carlo simulation. We have studied the magnetic properties and magnetocaloric effect in $Sr_2FeMoO_6$, $La_2SrMn_2O_7$ bilayer manganite, the surface effects on the magnetocaloric properties of perovskites ferromagnetic thin films, $SmFe_{1-x}Mn_xO_3$ perovskite.

# References

1. A. Farokhi, H. Shahroosvand, G. Delle Monache, M. Pilkington, M.K. Nazeeruddin, The evolution of triphenylamine hole transport materials for efficient perovskite solar cells, Chem. Soc. Rev. 51 (2022) 5974–6064.
2. P. Mahajan, B. Padha, S. Verma, V. Gupta, R. Datt, W.C. Tsoi, S. Satapathi, S. Arya, S, Review of current progress in hole-transporting materials for perovskite solar cells, J. Energy Chem. 68 (2022) 330–386.
3. X.X. Yin, Z.N. Song, Z.F. Li, W.H. Tang, toward ideal hole transport materials: a review on recent progress in dopant-free hole transport materials for fabricating efficient and stable perovskite solar cells, Energy Environ. Sci. 13 (2020) 4057–4086.
4. M. Ren, X. Qian, Y. Chen, T. Wang, Y. Zhao, Potential lead toxicity and leakage issues on lead halide perovskite photovoltaics, J. Hazard Mater. 426 (2022) 127848.
5. A. Babayigit, A. Ethirajan, M. Muller, B. Conings, Toxicity of organometal halide perovskite solar cells, Nat. Mater. 15 (2016) 247–251.
6. J.J. Yoo, G. Seo, M.R. Chua, T.G. Park, Y. Lu, F. Rotermund, Y.-K. Kim, C.S. Moon, N.J. Jeon, J.-P. Correa-Baena, V. Bulovi_c, S.S. Shin, M.G. Bawendi, J. Seo, Efficient perovskite solar cells via improved carrier management, Nature 590 (2021) 587–593, https://doi.org/10.1038/s41586-021-03285-w
7. Grancini G, Nazeeruddin MK. Dimensional tailoring of hybrid perovskites for photovoltaics. Nat Rev Mater 2019; 4:4–22.
8. Yuetian Chen, Xingtao Wang, Yao Wang, Xiaomin Liu, Yanfeng Miao, Yixin Zhao. Functional organic cation induced 3D-to-0D phase transformation and surface reconstruction of CsPbI3 inorganic perovskite. https://doi.org/10.1016/j.scib.2023.03.029
9. J. Miao, F. Zhang, Recent progress on highly sensitive perovskite photodetectors, J. Mater. Chem. C 7 (2019) 1741–1791.

10. Juan He, Xiaomin Xu, Meisheng Li, Shouyong Zhou, Wei Zhou. Recent advances in perovskite oxides for non-enzymatic electrochemical sensors: A review. Analytica Chimica Acta 1251 (2023) 341007

11. F. Malvano, L. Maritato, G. Carapella, P. Orgiani, R. Pilloton, M. Di Matteo, D. Albanese, Fabrication of SrTiO3 layer on Pt electrode for label-free capacitive biosensors, Biosensors 8 (2018) 26, https://doi.org/10.3390/bios8010026

12. U.T. Nakate, P. Patil, Y.T. Nakate, S.I. Na, Y.T. Yu, Y.B. Hahn, Ultrathin ternary metal oxide Bi2MoO6 nanosheets for high performance asymmetric supercapacitor and gas sensor applications, Appl. Surf. Sci. 551 (2021), 149422, https://doi.org/10.1016/j.apsusc.2021.149422

13. Y. Long, T. Saito, T. Tohyama, K. Oka, M. Azuma, Y. Shimakawa Inorganic Chemistry, 48 (2009), p. 8489

14. Y. Long, T. Kawakami, W. Chen, T. Saito, T. Watanuki, Y. Nakakura, Q. Liu, C. Jin, Y. Shimakawa Chem. Mater., 24 (2012), p. 2235

15. A. Aimi, D. Mori, K. Hiraki, T. Takahashi, Y.J. Shan, Y. Shirako, J. Zhou, Y. Inaguma Chem. Mater., 26 (2014), p. 2601

16. U. Dutta, A. Haque, M.M. Seikh, Synthesis, structure and magnetic properties of Ti doped La2MnNiO6 double perovskite, Chimica Techno Acta, 6 (2019) 80–92.

17. D.M.A. Jaimes, J.M. De Paoli, V. Nassif, P.G. Bercoff, G. Tirao, R.E. Carbonio, F. Pomiro, Effect of B-Site order-disorder in the structure and magnetism of the new perovskite family La2MnB'O6 with B' = Ti, Zr, and Hf, Inorg. Chem., 60 (2021) 423 4935–4944.

18. C.B. Samantaray, H. Sim, H. Hwang, Microelectron. J. 36 (2005) 725.

19. H. Muta, K. Kurosaki, S. Yamanaka, J. Alloys Compd. 350 (2003) 292.

20. N.O. Damerdji, B. Amrani, K.D. Khodja, P. Aubert, J. Supercond. Nov. Magnetism 31 (2018) 2935.

21. K. Xu, C. Lin, X. Xie, A. Meijerink, Chem. Mater. 29 (2017) 4265.

22. G. Pan, X. Bai, D. Yang, X. Chen, P. Jing, S. Qu, L. Zhang, D. Zhou, J. Zhu, W. Xu, B. Dong, H. Song, Nano Lett. 17 (2017) 8005.

23. K. Xu, A. Meijerink, Chem. Mater. 30 (2018) 5346.

24. Y.-Y. Sun, M.L. Agiorgousis, P. Zhang, S. Zhang, Chalcogenide Perovskites for Photovoltaics, Nano Lett. 15 (1) (2015) 581–585.

25. S. Perera, H. Hui, C. Zhao, H. Xue, F. Sun, C. Deng, N. Gross, C. Milleville, X. Xu, D. F. Watson, B. Weinstein, Y.-Y. Sun, S. Zhang, H. Zeng, Chalcogenide perovskites - an emerging class of ionic semiconductors, Nano Energy 22 (2016) 129–135.

26. X. Wei, et al., Realization of BaZrS3 chalcogenide perovskite thin films for optoelectronics, Nano Energy 68 (2020), 104317.

27. S. Niu, H. Huyan, Y. Liu, M. Yeung, K. Ye, L. Blankemeier, T. Orvis, D. Sarkar, D. J. Singh, R. Kapadia, J. Ravichandran, Bandgap Control via Structural and Chemical Tuning of Transition Metal Perovskite Chalcogenides, Adv. Mater. 29 (9) (2017) 1604733, https://doi.org/10.1002/adma.v29.910.1002/adma.201604733

28. I. Sadeghi, K. Ye, M. Xu, Y. Li, J.M. LeBeau, R. Jaramillo, Making BaZrS3 Chalcogenide Perovskite Thin Films by Molecular Beam Epitaxy, Adv. Funct. Mater. 31 (45) (2021) 2105563, https://doi.org/10.1002/adfm.v31.4510.1002/adfm.202105563

29. K. Hanzawa, S. Iimura, H. Hiramatsu, H. Hosono, Material design of green-lightemitting semiconductors: Perovskite-type sulfide SrHfS3, J. Am. Chem. Soc. 141 (13) (2019) 5343–5349

# Contents

**1 Calculation Methods: Monte Carlo Simulations and Ab Initio Calculations** ................................................... 1
  1.1 Introduction ................................................ 1
  1.2 Monte Carlo Method and Ising Model ........................ 2
  1.3 Ab-Initio Calculations ...................................... 7
  1.4 Several Approximations ..................................... 8
    1.4.1 Born–Oppenheimer Approximation .................... 8
    1.4.2 Hartree Approximation ............................. 9
    1.4.3 Hartree–Fock Approximation ......................... 10
  1.5 Density Functional Theory ................................... 11
    1.5.1 Theorems of Hohenberg and Kohn ................... 11
    1.5.2 Formulation of Kohn–Sham ......................... 12
    1.5.3 Locale Density of Approximation .................... 13
    1.5.4 Generalized Gradient Approximation ................. 14
  1.6 Conclusion ................................................ 15
  References .................................................... 15

**2 Magnetocaloric Effect, Electronic and Magnetic Properties in Manganite Perovskites** ...................................... 17
  2.1 Introduction ............................................... 17
  2.2 Calculation Details: Density-Functional Theory and Monte Carlo Simulations ......................................... 18
  2.3 Crystal Structure of Manganite Perovskite .................... 20
  2.4 Electronic Properties of Manganite Perovskite ................. 21
  2.5 Magnetic and Magnetocaloric Properties of Manganite Perovskite ................................................ 27
  2.6 Conclusions ............................................... 35
  References .................................................... 36

**3  Study of Magnetocaloric Effect, Electronic and Magnetic
     Properties of Perovskite Ferrites** ...................................  39
   3.1   Introduction  .................................................  39
   3.2   Density-Functional Theory and Monte Carlo Simulations  ........  40
   3.3   Crystal Structure of Ferrite Perovskite  .......................  43
   3.4   Electronic Properties of Ferrite Perovskite  ....................  44
   3.5   Magnetic and Magnetocaloric Properties of Ferrite Perovskites ....  49
   3.6   Conclusions  ..................................................  54
   References  ........................................................  56

**4  Magnetic and Magnetocaloric, Electronic, Magneto-Optical,
     and Thermoelectric Properties of Perovskite Chromites** ............  59
   4.1   Introduction  .................................................  59
   4.2   Calculation Methods  ..........................................  60
          4.2.1   Density Functional Theory  ...........................  60
          4.2.2   Monte Carlo Study  ..................................  60
   4.3   Crystal Structure of Perovskite Chromites  .....................  63
   4.4   Electronic Properties of Perovskite Chromites  ..................  64
   4.5   Dielectric and Optical Properties of Perovskite Chromites  ........  65
   4.6   Thermoelectric Properties of Perovskite Chromites  ..............  69
   4.7   Magnetic and Magnetocaloric Effect of Perovskite Chromites .....  71
   4.8   Conclusions  ..................................................  73
   References  ........................................................  74

**5  Magnetic Properties and Magnetocaloric in Double Perovskite
     Oxides**  .........................................................  77
   5.1   Introduction  .................................................  77
   5.2   Ising Model and Monte Carlo Simulations  .....................  78
   5.3   Results and Discussion of Magnetic Properties
          and Magnetocaloric in Double Perovskite Oxides  ..............  81
   5.4   Conclusions  ..................................................  82
   References  ........................................................  85

**6  Magnetocaloric and Magnetic Properties of Bilayer Manganite** .....  87
   6.1   Introduction  .................................................  87
   6.2   Ising Model and Monte Carlo Simulations  .....................  88
   6.3   Magnetic and Magnetocaloric Properties of Bilayer
          Manganite System .............................................  90
   6.4   Conclusions  ..................................................  92
   References  ........................................................  93

**7  Magnetocaloric Properties of Surface Effects in Perovskites
     Ferromagnetic Thin Films** .......................................  95
   7.1   Introduction  .................................................  95
   7.2   Ising Model and Monte Carlo Simulations  .....................  96
   7.3   Results and Discussion: Surface Effects in Perovskites
          Ferromagnetic Thin Films  ....................................  99

7.4   Conclusions .............................................. 107
References .................................................. 107

**8   Effect of Magnetic Field on the Magnetocaloric and Magnetic**
**Properties of Perovskite Orthoferrites** .......................... 109
8.1   Introduction .............................................. 109
8.2   Ising Model and Monte Carlo Study .......................... 110
8.3   Results and Discussion: Magnetocaloric Effect and Magnetic
Properties of Perovskite Orthoferrites ........................ 112
8.4   Conclusions .............................................. 116
References .................................................. 117

**General Conclusion** ............................................. 119

# About the Author

**Prof. Rachid Masrour** from Morocco, actually, is a research professor in the Faculty of Sciences Dhar El Mahraz at Sidi Mohamed Ben Abdellah the University, Fez, Morocco. Dr. Rachid Masrour completed his Ph.D. in March 2006 at the same University. His research interests lie in the areas of Condensed Matter Physics, Material Sciences, Material for energy, Magnetism, etc.... He has 340 articles published in Web of Science and 10 book chapters published in international journals, editor of 8 Books and author of 2 books with an h-index of 34. He participated in more than 100 Moroccan and international congress. He's a Referee of several articles and also an Editorial Board Member in different international journals. He has a lot of collaboration with laboratories from abroad. He has been honored with the International Association of Advanced Materials Young Scientist Medal in recognition for his contribution to "Magnetism, Electromagnetism and Spintronics" and delivered a lecture at the Advanced Materials World Congress on 11-14 October 2022. One of the world's most cited top scientists in material physics (top 2%, Stanford University Ranking, US, 2020 and in 2023). He has the Best Oral Presentation Award for MCGPD-2021, by Indian Association for Crystal Growth & Indian Science and Technology Association International Organization for Crystal Growth, 5-8, July 2021. Outstanding Scientist Award (VDGOOD PROFESSIONAL ASSOCIATION, 15-02-2020, India). It was among the most critical in 2020 by Elsevier.

# Chapter 1
# Calculation Methods: Monte Carlo Simulations and Ab Initio Calculations

**Abstract** In this chapter, I have given the description of Monte Carlo simulations and Ising model. The Metropolis algorithm has been detailed. The Ab-initio calculations have been given. The several approximations such as: Born–Oppenheimer Approximation, Hartree Approximation, Hartree–Fock Approximation, density functional theory, theorems of Hohenberg and Kohn, Formulation of Kohn–Sham, locale density of approximation and generalized gradient approximation were detailed in this chapter to solve the Schrodinger equation for N particles.

## 1.1 Introduction

Simulation methods have seen recent advancements across various domains, encompassing physical properties, pharmaceuticals, and biology. Modeling, which relies on computer-based techniques and approaches, has emerged as a means to study and analyze complex physical and mathematical problems, bypassing the need for experimental methods. Similarly, the growing application of theoretical calculations aids in deciphering experimental data that defies interpretation through traditional means, often due to intricate system complexities. Furthermore, theoretical calculations prove invaluable in forecasting outcomes and behaviors of compounds within challenging experimental conditions, such as extreme pressure or temperature scenarios. Computational methodologies efficiently facilitate system synthesis, data storage, retrieval, and the prediction of physical properties for materials, including those as elusive as nanomaterials. Our thesis will utilize quantum and classical calculations to ascertain magnetic properties and the magnetocaloric effect. These calculations will complement corresponding techniques, namely the Ab-initio approach and Monte Carlo simulation.

## 1.2    Monte Carlo Method and Ising Model

Monte Carlo simulations have a historical foundation predating the advent of computers, originating as experiments employing random numbers within a statistical framework. This versatile research tool finds extensive application across various professions, including medicine, biology, finance, operations research, and physics.

The significance of Monte Carlo simulations (MCS) lies in their substantial contributions to multiple fields in the past two decades. These simulations have become instrumental in problem-solving across disciplines like applied statistics, engineering, finance, business, design, computer science, telecommunications, and the physical sciences. Moreover, the evolution of MCS has underpinned the development of Bayesian multi-parameter problem-solving, leading to increased utilization of Bayesian statistics.

In this section, we will begin by providing an overview of various spin models in statistical physics, followed by a discussion of the Ising model. These spin models are categorized based on two critical considerations: degrees of freedom and atomic interactions. There are approximately three categories of spin models:

Models involving discrete spins, such as the well-known Ising model [1] and the Potts model [2].

Models involving continuous spins, including the XY dimensional unit vector model, the classical Heisenberg model with three-dimensional unit vectors [3], and N-dimensional unit vectors [4]. Additionally, there are models based on arrow configurations along network links, exemplified by Baxter's vertex models [5].

Within this thesis, we employ the Ising model to investigate both magnetic properties and the magnetocaloric effect of systems. Named after physicist Ernest Ising, this model serves as a mathematical representation of ferromagnetism in statistical mechanics. It consists of discrete variables representing atomic spins' magnetic dipole moments, which can assume two states ($+1$ or $-1$). These spins are arranged on a lattice, typically in the form of a graph, allowing for interactions with neighboring spins. The Ising two-dimensional square lattice model represents one of the simplest statistical models for illustrating a phase transition. The Hamiltonian Ising model is expressed as follows:

$$\mathcal{H} = - \sum_{<i,j>} J_{i,j} S_i S_j - H \sum_{i=1} S_i$$

In the equation you've provided, where $<i, j>$ represents the first nearest neighbors between spins $S_i - S_j$, and $J_{ij}$ denotes the exchange interaction between spins $S_i - S_j$, the next crucial step, once the physical system model has been selected, is to ascertain the model's statistical properties. To initiate this process, we begin by considering the probability of a configuration as defined by the Boltzmann distribution [6, 7]:

$$P(S_1, S_2, S_3, \ldots, S_N) = \frac{1}{Z} e^{-\beta \mathcal{H}(S_1, S_2, S_3, \ldots, S_N)}$$

where $\beta = \frac{1}{k_B T}$ s inverse of the temperature rescaled by the Boltzmann constant kB and Z refers to the partition function and given by:

$$Z = \sum_i \exp(-\beta E_i)$$

$E_i$ is the energy of state i and $k_B T$ is the thermal energy.

One of the primary goals in Monte Carlo simulations is to determine average values. To enhance the efficiency of this process, the results should be biased towards more probable values. The thermal mean of a quantity, Q, is calculated by summing the contributions from all possible system states, weighted by their respective probabilities:

$$<Q> = \frac{\sum_a Q_a e^{-\beta E_a}}{\sum_a e^{-\beta E_a}}$$

Calculating this average becomes computationally challenging for larger systems, as summing over all states can introduce inaccuracies. To address this, the Monte Carlo method randomly selects a subset of states from a specified distribution ($p_a$). For a finite number of states, $N = \{a_1, \ldots, a_M\}$, the corresponding probability distribution P(m) is needed:

$$Q_N = \frac{\sum_{i=1}^{N} Q_{a_i} p_{a_i}^{-1} e^{-\beta E_{a_i}}}{\sum_{i=1}^{N} p_{a_i}^{-1} e^{-\beta E_{a_i}}}$$

Thus, when $N \rightarrow \infty$, $<Q_N> = Q$ and N remains to be determined for the improved expression of Q. For this, equal probability is considered between system states, i.e. $p_{ai} = p_{aj}$.

$$Q_N = \frac{\sum_{i=1}^{N} Q_{a_i} e^{-\beta E_{a_i}}}{\sum_{i=1}^{N} e^{-\beta E_{a_i}}}$$

The challenging aspect lies in determining the suitable estimator within Monte Carlo simulations. Monte Carlo methods leverage this Markov process to decide which states to include.

The evolution of the Markov chain is described by the equation:

$$\frac{dP(S_a, t)}{dt} = -\sum_a w(a \rightarrow b) P_a(t-1) + \sum_b w(b \rightarrow a) P_b(t-1)$$

The transition probabilities, denoted as w(a → b) and w(b → a), represent the likelihood of transitioning from state a to b and vice versa, and t signifies the time elapsed in the Markov process. The first term encompasses all possible transitions to state a, while the second term encompasses all possible transitions to state b.

The generated state may not remain the same; it explores new states based on the transition probability w(a → b). Two conditions guide this search:

The transition probability remains constant over time.

The transition probability is contingent on the properties of the system in states a and b.

This implies that the transition probability, W(a → b), from one state a to another state b in the Markov process remains constant and must adhere to the relation:

$$\sum_a w(a \to b) = 1$$

Ergodicity represents a crucial condition in which the system can transition from one state to another over a sufficiently extended period during the Markov process. If all transition probabilities for a particular state are zero, ergodicity is not met.

The detailed balance condition signifies that the system remains in equilibrium, like how atoms leaving equilibrium are replaced by an equivalent number of atoms also departing equilibrium. Consequently, the Boltzmann distribution's probability exceeds that of all other distributions.

The usefulness of the detailed balance equilibrium is expressed mathematically as follows:

$$\sum_a P_a w(a \to b) = \sum_b P_b w(b \to a)$$

we can obtain:

$$P_a = \sum_b P_b w(b \to a)$$

Nevertheless, the detailed balance condition alone is insufficient to conclude that equilibrium is described by the Boltzmann–Gibbs distribution. To address this limitation, we introduce another condition on the transition probability. This additional condition, known as specific balance, is defined as follows:

$$P_a w(a \to b) = P_b w(b \to a)$$

This condition demonstrates that the specific equilibrium requirement rules out the concept of a limit cycle. Given that we are examining a system in thermal equilibrium, the probability distribution conforms to the Boltzmann equation, which can be expressed as follows:

$$w(a \to b)e^{-\beta E_a} = w(b \to a)e^{-\beta E_b}$$

Existing standard methods may occasionally fall short in addressing the unique requirements of each new problem. In light of this, we find it imperative to develop algorithms tailored to solve these challenges. In our approach to problem-solving, we propose several Markov processes. It's worth noting that when predicting the Markov process that yields the correct transition outcome becomes challenging, we employ the acceptance rate as a means to identify favorable transition probabilities from various Markov processes.

The transition probability is determined as follows:

$$w(a \to b) = g(a \to b)A(a \to b)$$

$g(a \to b)$ represents the selection probability, which involves generating a new state b from the current state a using an algorithm.

$(a \to b)$ is the acceptance probability, determining whether to accept the state change. The value of the acceptance probability, also referred to as "acceptance," is a random number ranging from 0 to 1.

If $(a \to b)$ equals 0 for all transitions, this effectively implies the following:

$(a \to b) = 0$ for all transitions implies that no state changes are accepted. In other words, the system remains in its current state without any transitions, and the Markov process effectively halts.

$$w(a \to a) = 1$$

$$\frac{w(a \to b)}{w(b \to a)} = \frac{g(a \to b)A(a \to b)}{g(b \to a)A(b \to a)}$$

$$\text{where } \frac{A(a \to b)}{A(b \to a)} \epsilon [0, \infty[$$

The terms $g(a \to b)$ and $g(b \to a)$ can take on various values. To ensure the algorithm's efficiency, the acceptance rate is typically chosen to be close to 1. In the development of the Monte Carlo algorithm, we will create a process that generates successive states primarily using the data from $g(b \to a)$. Subsequently, we will select the relevant states based on the acceptance probability.

In the upcoming section, we will delve into the Metropolis algorithm. This algorithm, initially devised in 1953 by Nicholas Metropolis and his colleagues at the Los Alamos laboratory in Mexico [8], was originally designed for calculating equations of state in molecular mixtures and their interactions. The central component of this algorithm is a Markov chain: starting from an initial state (a) characterized by the Boltzmann factor $\exp(-\beta E_a)$, it involves transitioning a particle from one state (a) to a new state (b) with a Boltzmann factor $\exp(-\beta E_b)$. This transition is carried out using the transition probability $W(a \to b)$ from state (a) to state (b). Each state belongs to a finite group of states known as the "state space."

The Metropolis algorithm operates as follows (Fig. 1.1):

Randomly select a spin from the set of N spins and record it.

Compute the energy associated with this new state and calculate the transition probability.

**Fig. 1.1** Flowchart of the
Metropolis algorithm [9]

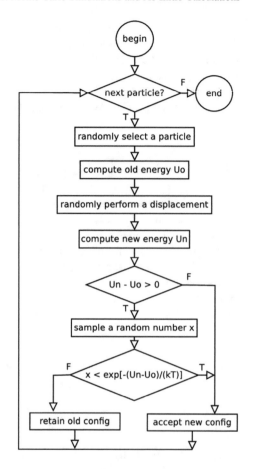

$$W(a \rightarrow b) = \exp(-\Delta E/k_B T).$$

To determine whether the flip is accepted or rejected, a random number, denoted as "r," is uniformly selected from the range between 0 and 1.

If the transition probability $W(a \rightarrow b)$ is greater than the random number r, then the flip is accepted; otherwise, it is rejected. This step is referred to as the "detailed equilibrium condition."

Subsequently, update the values of thermodynamic quantities, with particular attention to the energy, and record the data of interest.

Return to the random selection of a spin to flip and repeat steps 1–6 until a sufficiently significant number of statistically relevant samples have been obtained.

The Metropolis algorithm's iterative nature, based on these steps, enables the exploration and sampling of the system's state space to calculate properties of interest.

## 1.3  Ab-Initio Calculations

The primary objective of the Ab-initio method is to predict material properties encompassing magnetic, electronic, and optical characteristics. These properties are determined by solving the quantum mechanical equations without relying on adjustable variables. To tackle this equation, density functional theory (DFT) has emerged as a valuable tool [10]. DFT is grounded in the electron density and has become one of the most widely utilized methods for quantum calculations concerning electronic structures in solid-state materials. Moreover, DFT empowers scientists to compute the ground state of systems comprising numerous electrons, making it the method of choice for investigating the physical properties of solids' ground states.

Material properties can be investigated by solving the time-independent Schrödinger equation, which is represented as follows:

$$\widehat{H}\Psi = E\Psi.$$

where:

$\hat{H}$ represents the Hamiltonian operator of a system.

$\Psi$ symbolizes the wave function, which describes the quantum state of the system.

E denotes the total energy of the system.

This equation serves as a fundamental framework for understanding the quantum behavior of particles in a physical system, providing insights into their energy levels and wave functions. Solving this equation plays a pivotal role in quantum mechanics and allows us to comprehend and predict the behavior of particles and systems at the atomic and subatomic scales.

The total Hamiltonian $\widehat{H}$, associated with a system with many interacting particles (n electrons + N nucleus), is given by the sum of the total kinetic energy operator $\widehat{T}$ and by the sum of the coulombic interactions operator $\widehat{V}$ (energy potential). Thus, it can be formulated in the following expression [11]:

$$\widehat{H}_T = \widehat{T}_N + \widehat{T}_e + \widehat{V}_{N-e} + \widehat{V}_{N-N} + \widehat{V}_{e-e}$$

with

$\widehat{T}_N = -\frac{\hbar^2}{2}\sum_i \frac{\nabla_{R_i}^2}{M_N}$: is the kinetic energy of the nucleus of mass MN.

$\widehat{T}_e = -\frac{\hbar^2}{2}\sum_i \frac{\nabla_{r_i}^2}{m_e}$: refers to the kinetic energy of electrons of mass me.

$\widehat{V}_{N-e} = -\frac{1}{4\pi\varepsilon_0}\sum_i \frac{e^2 Z_i}{|R_i - r_j|}$: corresponds to the potential energy of the nucleus-electron interaction.

$\widehat{V}_{e-e} = \frac{1}{8\pi\varepsilon_0}\sum_{i\neq j} \frac{e^2}{|r_i - r_j|}$: corresponds to the electron–electron repulsion potential energy.

$\widehat{V}_{N-N} = \frac{1}{8\pi\varepsilon_0}\sum_{i\neq j}\frac{e^2 Z_i Z_j}{|R_i - R_j|}$: corresponds to the core-nucleus repulsion potential energy.

Thus, we can rewrite the total Hamiltonian as follows:

$$\widehat{H}_T = -\frac{\hbar^2}{2}\sum_i\frac{\nabla_{R_i}^2}{M_N} - \frac{\hbar^2}{2}\sum_i\frac{\nabla_{r_i}^2}{m_e} - \frac{1}{4\pi\varepsilon_0}\sum_{i,j}\frac{e^2 Z_i}{|R_i - r_j|}$$

$$+ \frac{1}{8\pi\varepsilon_0}\sum_{i\neq j}\frac{e^2 Z_i Z_j}{|R_i - R_j|} + \frac{1}{8\pi\varepsilon_0}\sum_{i\neq j}\frac{e^2}{|r_i - r_j|}$$

Indeed, the analytical resolution of the total Hamiltonian for a multi-particle system, comprising N nuclei and n electrons interacting and moving within an electromagnetic field, is exceedingly challenging and often impossible due to the complexity of the interactions involved. Consequently, it becomes essential to resort to various approximations to tackle such systems. Some of the most prominent approximations used in quantum chemistry and physics.

## 1.4 Several Approximations

### 1.4.1 Born–Oppenheimer Approximation

The first major approximation, introduced by Born and Oppenheimer in 1927 [12], is known as the Born–Oppenheimer approximation. This approximation involves separating the motions of the electrons from those of the nuclei within a molecule or system. The rationale behind this separation is that electrons are significantly lighter than nuclei, and as a result, nuclei move much more slowly in comparison to electrons. Thus, it is reasonable to assume that the positions of the nuclei remain essentially fixed during electronic motion.

Under the Born–Oppenheimer approximation, the total Hamiltonian of the system becomes a Hamiltonian that depends solely on the electronic coordinates:

$$\widehat{H}_T = \widehat{T}_e + \widehat{V}_{e-e} + \widehat{V}_{N-e}$$

$$\text{avec } \widehat{V}_{N-e} = \widehat{V}_{ext}$$

Thus, we can write the Hamiltonian as follows:

$$\widehat{H}_T = -\frac{\hbar^2}{2}\sum_i\frac{\nabla_{r_i}^2}{m_e} + \frac{1}{8\pi\varepsilon_0}\sum_{i\neq j}\frac{e^2}{|r_i - r_j|} - \frac{1}{4\pi\varepsilon_0}\sum_{i,j}\frac{e^2 Z_i}{|R_i - r_j|}$$

Indeed, while the Born–Oppenheimer approximation simplifies the Schrödinger equation significantly by separating the electronic and nuclear motions, it does not completely resolve the challenges posed by the complex electron–electron interactions within a system. As a result, other approximations, such as the Hartree approximation, Hartree–Fock approximation, and Density Functional Theory (DFT), have been proposed to further address these complexities and make quantum calculations feasible for a wide range of systems. These additional approximations provide various levels of sophistication and computational efficiency to tackle the intricacies of multi-electron systems and have been instrumental in advancing our understanding of molecular and materials properties. $\hbar$ is the Planck constant reduced.

### 1.4.2 Hartree Approximation

The attempts to address the many-electron problem in quantum mechanics are investigated by Ref. [13]. This approximation is rooted in the assumption of electron independence, meaning that each electron is considered to move in an average field created by the nuclei and all other electrons. Under the Hartree approximation, the total wave function ($\Psi$) for a multi-electron system is approximated as the product of the individual one-electron wave functions ($\psi_i$):

$$\Psi(r_1, r_2, \ldots, r_N) = \psi_1(r_1)\psi_2(r_2)\ldots\psi_N(r_N)$$

We obtain the Schrödinger equation for each wave function $\psi i$:

$$-\frac{\hbar^2}{2m}\Delta\psi_i(\vec{r}) + V_{eff}\psi_i(\vec{r}) = \varepsilon_i\psi_i(\vec{r})$$

The Hartree equations, denoted as $\varepsilon i$, indeed represent the electronic energy levels associated with a specific electronic state or orbital. The first term is the electron's kinetic energy, and the second term is the effective potential ($V_{eff}$) experienced by the electron under the influence of the nucleus as well as the other electron.

The potential of the electron-nucleus interaction can be described by:

$$V_{N-e}(\vec{r}) = -Ze^2 \sum_R \frac{1}{\left|\vec{r} - \vec{R}\right|}$$

For electron–electron interaction, the electron is considered to be moving in a Hartree potential expressed by:

$$V_H(\vec{r}) = -e \sum_r \frac{d\vec{r'}}{\left|\vec{r} - \vec{r'}\right|}\rho(\vec{r'})$$

The effective potential can be expressed as follows:

$$V_{eff}(\vec{r}) = V_H(\vec{r}) + V_N(\vec{r})$$

We now turn to the Hartree–Fock approach, which adds a further correction to the Hartree approximation.

### 1.4.3  Hartree–Fock Approximation

Certainly, the Hartree–Fock approximation, as discussed in Ref. [14], relies significantly on the Pauli exclusion principle. This principle, a fundamental concept in quantum physics, applies to fermions, which are particles characterized by half-integer spin values, including electrons, protons, and neutrons. The Pauli exclusion principle states that two or more identical fermions cannot occupy the same quantum state within the same spatial region simultaneously. This leads us to consider the antisymmetric of the wave function as follows:

$$\Psi(r_1, \ldots, r_a, \ldots, r_b, \ldots, r_N) = -\Psi(r_1, \ldots, r_a, \ldots, r_b, \ldots, r_N)$$

Antisymmetric is introduced into the Schrödinger equation. The new wave function is described by the Slater determinant [15]. We obtain the following Schrödinger equation for an electron:

$$-\frac{\hbar^2}{2m}\Delta\Psi_i(\vec{r}) + V_{eff}\Psi_i(\vec{r}) - \sum_j \left\{ \int \frac{d^3\vec{r}'}{\left|\vec{r} - \vec{r}'\right|} \Psi_j^*\left(\vec{r}'\right)\Psi_i\left(\vec{r}'\right) \right\} \Psi_i(\vec{r})$$
$$= \varepsilon_i\psi_i(\vec{r})$$

The key difference indeed lies in the introduction of the exchange term in the Hartree–Fock method. This exchange term considers the antisymmetric of the wave function, as required by the Pauli exclusion principle. As a result, the Hartree–Fock approximation offers a more accurate description of electronic structure by addressing electron–electron correlations that are not accounted for in the simpler Hartree approximation. However, this additional complexity makes the Hartree–Fock method more challenging to solve computationally.

## 1.5 Density Functional Theory

DFT, initially developed by Hohenberg [16], Kohn, and Sham, serves as a powerful tool for electronic structure calculations in the fields of solid-state physics and quantum chemistry. The fundamental concept behind DFT is that the electronic properties of a system can be determined primarily from the electron density, denoted as $\rho(\vec{r})$. By focusing on the electron density, DFT formulates the problem of solving the Schrödinger equation using various theorems and approaches.

Kohn and Sham's theorem, a pivotal contribution to DFT, simplified the problem of solving the electronic structure problem. This theorem builds upon the work of Thomas and Fermi [17, 18], who initially treated the system as a homogeneous electron gas and considered its kinetic energy as a local density functional. However, they neglected electron–electron interactions and exchange–correlation effects.

To address the shortcomings of the original Thomas–Fermi approach, Dirac introduced exchange–correlation effects through local exchange corrections [19]. These corrections account for the interactions and correlations among electrons, thus providing a more accurate foundation for DFT calculations.

### 1.5.1 Theorems of Hohenberg and Kohn

For a system of n interacting electrons, the Hamiltonian can be written as:

$$\widehat{H} = \widehat{T} + \widehat{V}_{ee} + \widehat{V}_{ext}$$

where, $\widehat{T}$ denotes kinetic energy, $\widehat{V}_{ee}$ describing interelectronic repulsions and $\widehat{V}_{ext}$ describing the external potential due to the nuclei.

Hohenberg and Kohn proved that for such a system, the ground-state wave function is a unique function of the electron density, $\Psi_0 = \Psi(\rho_0)$. Therefore, any observable $\widehat{O}$ depends on the functional density. Thus, the total energy of the system is written as follows:

$$E_V(\rho) = F_{HK}(\rho) + V_{ext}(\rho)$$

with $F_{HK}(\rho) = T(\rho) + V_{ee}(\rho)$ et $V_{ext}(\rho) = \int \widehat{V}_{ext}(r)\rho(r)dr$ is the universal function of Hohenberg and Kohn and the external potential, respectively.

The total energy of the system $E_v(\rho)$ is minimal when the density $\rho(\vec{r})$ corresponds to the electron density $(\vec{r})$ of the ground state.

The Hohenberg–Kohn theorems pertain to systems comprising electrons moving within an external potential, and they can be summarized as follows:

**First Hohenberg–Kohn Theorem**: The external potential ($V_{ext}$) uniquely determines the ground-state electron density ($\rho(\vec{r})$) of the system. In other words, for a given external potential, there is a one-to-one correspondence between the external potential and the ground-state electron density.

**Second Hohenberg–Kohn Theorem**: The ground-state energy (E0) of the system is a unique functional of the ground-state electron density ($E_0[\rho(\vec{r})]$). This means that the ground-state energy can be expressed solely as a functional of the electron density and is minimized when the true ground-state electron density is used.

For any system of interacting particles in an external potential $\widehat{V}_{ext}(r)$, this potential is determined solely, except for one constant, by the density of the ground state $\rho_0(\vec{r})$.

A universal energy function $E(\rho)$ in terms of density $\widehat{V}_{ext}(r)$. The exact ground-state energy of the system is the global minimum of this functional, and the density that minimizes the functional is the exact ground-state density $\rho_0(\vec{r})$.

Knowledge of the function $F_{HK}[\rho(\vec{r})]$ is required to accurately determine the total energy and ground-state properties of the system. However, this function $F_{HK}[\rho(\vec{r})]$ is difficult to determine precisely. It is therefore necessary to use the Kohn–Sham formulation, which is described in what follows.

$V_{ext}(\rho)$ is the universal function of (H.K) and the external potential, respectively.

The total energy of the system $E_v(\rho)$ is minimal when the density $\rho(\vec{r})$ corresponds to the electron density $\rho_0(\vec{r})$ of the ground state.

In brief, Hohenberg and Kohn's theorems, related to any system consisting of electrons moving under the influence of the external potential, can be stated as follows:

**Theorem 1.1** *For any system of interacting particles in an external potential $\widehat{V}_{ext}(r)$, this potential is determined solely, except for one constant, by the ground-state density $\rho_0(\vec{r})$.*

**Theorem 1.2** *A universal energy function $E(\rho)$ in terms of density $\rho(\vec{r})$ can be defined, and valid for any external potential $\widehat{V}_{ext}(r)$. The exact ground-state energy of the system is the global minimum of this functional, and the density that minimizes the functional is the exact ground-state density $\rho_0(\vec{r})$.*

Knowledge of the $F_{HK}[\rho(\vec{r})]$ is necessary to accurately determine the total energy and ground-state properties of the system. However, this $F_{HK}[\rho(\vec{r})]$ is difficult to determine accurately. It is therefore necessary to use the Kohn–Sham formulation, which is described in what follows.

### 1.5.2 Formulation of Kohn–Sham

The Kohn–Sham formulation is a fundamental concept in density functional theory (DFT). It replaces the complex interacting system of electrons with a simpler, non-interacting auxiliary system while maintaining the same particle number density as the interacting system. This approach allows for more tractable calculations of electronic properties.

In this context, the energy of the system (consisting of non-interacting electrons) can be expressed in the following form:

$$E_V(\rho) = T_s(\rho) + V_{ext}(\rho) + V_e(\rho) + E_{xc}(\rho)$$

with $T_s(\rho)$: describes the kinetic energy of non-interacting electrons.

$V_{ext}(\rho)$: is the Coulomb energy of the electron–electron interaction.

$E_{xc}(\rho)$: corresponds to the exchange correlation energy.

$T_s(\rho)$ can be written as follows: $T_s(\rho) = \sum_{i=1}^{N} \langle \Psi_i | -\frac{\nabla^2}{2} | \Psi_i \rangle$

with $\rho(r) = \sum_{i=1}^{N} |\Psi_i(r)|^2$.

By minimizing $E_V(\rho)$ on $\Psi_i$ and considering H.K alongside the introduction of Lagrange($\varepsilon_i$) parameters, the equation is as follows:

$$\frac{\partial}{\partial \Psi_i^*(r)} \left[ E_v - \varepsilon_i \int var \Psi_i^*(r)\Psi_i(r)dr \right]$$

After a series of calculations, the Kohn–Sham equations are given by:

$$\left[ -\frac{\nabla^2}{2} + V_{eff}(r) \right] \Psi_i(r) = \varepsilon_i \Psi_i(r)$$

$$Ou \ V_{eff}(r) = V_{ext}(r) + V_H(r) + V_{xc}(r)$$

$V_H(r) = \int \frac{\rho(r')}{|r-r'|} dr$ and $V_{xc}(r)$ show the Hartree potential and exchange correlation potential respectively.

The development of the Kohn–Sham equations indeed underscores that, within this formalism, the only remaining unknown is the exchange–correlation functional, denoted as $E_{xc}[\rho((\vec{r}))]$. This functional encapsulates the electron–electron exchange and correlation effects and is a crucial component of density functional theory (DFT).

### 1.5.3  Locale Density of Approximation

While the LDA is relatively simple and computationally efficient, it may not capture certain electron–electron correlation effects accurately, especially in systems with strong electron localization or in molecules with significant charge density variations. Therefore, more advanced approximations, such as the Generalized Gradient Approximation (GGA) and hybrid functionals, have been developed to improve upon the limitations of the LDA and provide more accurate descriptions of electronic structure and properties in a wider range of systems.

The density $\rho(\vec{r})$ varies slowly with position $(\vec{r})$,

The contribution of the exchange correlation energy $E_{xc}[\rho(\vec{r})]$ to the total system energy can be added cumulatively from each portion of the non-uniform gas as if it were locally uniform.

Using this approximation, the exchange correlation energy for a density $\rho(\vec{r})$ is given by:

$$E_{xc}^{LDA}[\rho(\vec{r})] \approx \int \epsilon_{xc}^{LDA}[\rho(\vec{r})]\rho(\vec{r})d\vec{r}$$

with $E_{xc}^{LDA}[\rho(\vec{r})]$: Exchange correlation energy.

The exchange correlation potential can also be written as a function of $\epsilon_{xc}^{LDA}[\rho]$ as shown in equation:

$$V_{xc}^{LDA}(\vec{r}) = \frac{\partial(\rho(\vec{r})\epsilon_{xc}^{LDA}[\rho(\vec{r})])}{\partial\rho(\vec{r})}$$

It has been shown that this function cannot study all systems whose electron density is highly variable with position $(\vec{r})$. Thus, we use another approximation.

### 1.5.4   Generalized Gradient Approximation

GGA functionals have been developed to provide better accuracy in describing various types of chemical bonds, charge density distributions, and other electronic properties. They are particularly valuable for systems with spatial variations in electron density, such as molecules with complex structures or materials with heterogeneous electron distributions.

Overall, GGA has become a widely used and effective tool in DFT, providing more accurate predictions of electronic properties in a diverse range of systems compared to the simpler LDA approximation, $E_{xc}^{GGA}\rho(\vec{r})$ but also of the electron density gradient $\nabla\rho(\vec{r})$.

The GGA functional can be written as follows [20]:

$$E_{xc}^{GGA}[\rho(\vec{r})] \approx \int \epsilon_{xc}^{GGA}[\rho(\vec{r}), |\nabla[\rho(\vec{r})]|]\rho(\vec{r})d\vec{r}$$

In this new formalism, the exchange correlation energy term per particle in an inhomogeneous electronic system is given by: $\epsilon_{xc}^{GGA}[\rho(\vec{r}), |\nabla[\rho(\vec{r})]|]$. The development of these methods has significantly expanded the applicability of DFT to systems involving van der Waals forces, including complex molecular structures, layered materials, and adsorption phenomena. Researchers now have a range of tools within DFT, including various functionals and correction schemes, to choose from depending on the specific characteristics of the system they are studying. This has contributed to the continued success and widespread use of DFT in a diverse array of scientific disciplines.

## 1.6 Conclusion

Both Monte Carlo simulations and Ab-initio simulations have found extensive use in scientific research and computational studies. They offer complementary approaches to understanding and predicting the behavior of physical and mathematical systems, making them valuable tools in various scientific disciplines.

## References

1. W.E. Pickett, Pseudopotential methods in condensed matter applications. Comput. Phys. Rep. **9**, 115–197 (1989)
2. E.Z. Ising, Physics **31**, 253 (1925)
3. R.B. Potts, Proc. Camb. Philos. Soc. **48**, 106 (1952)
4. M. Yeomans, *Statistical Mechanics of Phase Transitions* (Oxford University Press, Oxford, 1993)
5. H.E. Stanley, Phys. Rev. Lett. **20**, 589 (1968)
6. L. Onsager, Phys. Rev. **65**, 117 (1944)
7. D. Landau, K. Binder, *A Guide to Monte Carlo Simulations in Statistical Physics* (Cambridge University Press, 2000)
8. N. Metropolis, A. Rosenbluth, M. Rosenbluth, A. Teller, E. Teller, J. Chem. Phys. **21**, 1087 (1953)
9. L. Muccioli et al., Top. Curr. Chem. **352**, 39–102 (2014)
10. L.H. Thomas, Proc. Camb. Philos. Soc. **23**, 542 (1927)
11. M. Attarian Shandiz, Monte Carlo and Density Functional Theory Simulation of Electron Energy Loss Spectra, Thesis, McGill University, Canada (2014)
12. M. Born, J.R. Oppenheimer, Ann. Phys. **87**, 457 (1927)
13. D.R. Hrtree, Proc. Camb. Philos. Soc. **24**, 89 (1928)
14. I. Koutiri, "étude ab-initio du trioxyde de tungstène wo3 en volume et en surface", these, université de Montpellier II sciences et techniques du Languedoc, Montpellier (2012)
15. P. Kiréev, *La physique des semiconducteurs,* 2e édition (Mir, Moscow, 1975)
16. P.C. Hohenberg, W. Kohn, Phys. Rev. B **136**, 864 (1964)
17. W. Kohn, L.J. Sham, Self-consistent equations including exchange and correlation effects. Phys. Rev. A **140**, 1133 (1965)
18. L.H. Thomas, The calculation of atomic fields. Proc. Camb. Philos. R. Soc. **23**, 542–548 (1927)
19. E. Fermi, Un metodostatistico per la determinazione dialcunepriorietadell "atome". Rend. Accad. Naz. Lincei **6**, 602–607 (1927)
20. P.A.M. Dirac, Note on exchange phenomena in the Thomas-Fermi atom. Proc. Camb. Philos. R. Soc. **26**, 376–385 (1930)

# Chapter 2
# Magnetocaloric Effect, Electronic and Magnetic Properties in Manganite Perovskites

**Abstract** The study conducted on $Pr_{0.65}Sr_{0.35}MnO_3$ perovskite involved the utilization of first principal calculations and Monte Carlo simulations to investigate its magnetocaloric effect, electronic properties, and magnetic properties. First, the electronic and magnetic properties were determined by treating the exchange–correlation potential with the generalized gradient approximation. It was found that $Pr_{0.65}Sr_{0.35}MnO_3$ perovskite exhibits ferromagnetic behaviour and possesses a half-metallic nature with 100% spin polarization. This characteristic is significant in relation to the compound's colossal magnetoresistance properties. Furthermore, the magnetic moment and the thermal variation of magnetization of $Pr_{0.65}Sr_{0.35}MnO_3$ perovskite were obtained through the first principal calculations. These calculations provide valuable insights into the material's magnetic behaviour and how it changes with temperature. To study the temperature dependence of the magnetic entropy change and the adiabatic temperature, Monte Carlo simulations were employed. These simulations allow for the investigation of the thermodynamic properties of the material, such as its response to changes in temperature and magnetic field. Based on the Monte Carlo simulations, the Curie temperature, which represents the temperature at which the material transitions from a ferromagnetic to a paramagnetic state, was deduced. Additionally, the field dependence of the relative cooling power, a measure of the material's efficiency in magnetic refrigeration applications, was determined. In summary, the combination of first principal calculations and Monte Carlo simulations provided valuable insights into the magnetocaloric effect, electronic properties, and magnetic properties of $Pr_{0.65}Sr_{0.35}MnO_3$ perovskite, including its half-metallic nature, Curie temperature, and field dependence of relative cooling power.

## 2.1 Introduction

The Density Functional Theory (DFT) method is used to interpret experimental observations and predict new material properties [1, 2]. In addition, the DFT method makes it possible to model systems that are closer or identical to those elaborate in laboratory. On the other hand, to understand physical phenomena such as colossal magnetoresistance and magnetocaloric effect, the study of electronic properties is necessary. Therefore, it gives us a global view of the physical properties of materials in the ground state. So far, the DFT method is the only method that accurately addresses these electronic properties [3, 4]. Perovskite oxide is a kind of material which has great magnetocaloric effect at room temperature [5, 6]. On the other hand, manganese-based perovskites generate important relationships between electronic properties and magnetic [7–14]. The substitution of an atom belonging to the lanthanide group by a divalent cation leads to the partial conversion of manganese ions $Mn^{3+}$ into $Mn^{4+}$ which results in a progressive weakening of ferromagnetism as the rate of $A^{3+}$ ion doping increases [15–17]. In addition, other research work focused on the effect of substitution of Pr and La by Sr showed an improvement in the magnetocaloric effect. In particular, the doping of the Pr and La atom by 0.35 and 0.25% of Sr respectively has been widely studied [18–21]. The magnetic, electronic, and structural properties of these compounds are strongly dependent on the valence state, the spin state of the metal ions and the defects. In this chapter, we resumed the study of $(La, Pr)_{1-x}Sr_xMnO_3$ with the aim of showing the influence of the substitution of Sr by Pr and La on the electronic, magnetic and magnetocaloric properties in these manganese-based. We relied on two methods; the first method is called DFT theory. This theory also helps us determine the values of Metropolis Monte Carlo algorithm input parameters. In particular, the moments and couplings magnetic. The second method is Monte Carlo simulations.

## 2.2   Calculation Details: Density-Functional Theory and Monte Carlo Simulations

To study the physical properties of matter, there are many computational methods that are becoming more and more perfect have been put at the service of researchers. Among which are used for the potential, the charge density and the wave base on which the wave functions are investigated. LAPW method was developed by Andersen, is basically an improvement of the so-called Augmented Plane Wave method developed by Slater [22, 23].

In this chapter, we used the Linearized Augmented Plane Wave method in the framework of the density functional theory (DFT) implemented in the WIEN2k code [24]. Exchange and correlation energy are examined using the gradient generalized approximation GGA. Kohn–Sham wave functions were expanded as combinations of spherical harmonic functions inside non-overlapping spheres around the atomic sites and in Fourier series in the interstitial region. To guarantee accurate results in a relatively reasonable time, the choice of input parameters must be optimal. For some parameters, this choice was made by a convergence test, while for others, it was based on a value close to the limit value. For this method, the choice of muffin-tin radii is important for the different atoms that form the different perovskites chosen. Muffin-tin radii represents the boundary between two regions of space, where potentials and wave functions are constructed in ways different.

These parameters are manifested by: The energy of separation between the valence and core states for $La_{0.75}Sr_{0.25}MnO_3$ and $Pr_{0.65}Sr_{0.35}MnO_3$ is $-6.0$ and $-9.0$ $R_y$, respectively. The valence wave functions inside the muffin-tin spheres are expanded in terms of spherical harmonics up to $l_{max} = 10$. The wave functions in the interstitial region were expanded in plane waves with a cutoff of $R_{MT}$ (where $R_{MT}$ is the average radius of the MT spheres). The muffin-tin radius RMT is based on two conditions: (i) no core charge leaks out of MT spheres and (ii) no overlapping is permitted between spheres. The muffin-tin radii of La, Sr, Mn and O were chosen to be 2.41, 2.20, 2.01 and 1.73 respectively for $La_{0.75}Sr_{0.25}MnO_3$. While in $Pr_{0.65}Sr_{0.35}MnO_3$ the muffin-tin radii of Pr, Sr, Mn and O were chosen to be 2.45, 2.23, 1.95 and 1.68, respectively.

The Monte Carlo simulations is based on the Metropolis algorithm that imposed cyclic boundary conditions on the network [25]. We generated setups by sequentially traversing the network and attempting single-spin flips. The spin flip is chosen according to the energy difference between the considered spin and its neighbors. When the energy difference is negative, the spin will reverse directly. Otherwise, it will flip if the randomly chosen number is less than the Boltzmann factor $e^{-\beta E}$.

The Hamiltonian of these systems is:

$$H = - \sum_{<i,j>} J_{ij} S_i S_j - H \sum_i S_i$$

where $S_i$ and $S_j$ respectively indicate the spin at the lattice site $i$ and site $j$. H is the magnetic field. The spin moment of $Mn^{3+}$ is $S = 2$ in both compounds. The summation runs over all pairs of nearest—, next-nearest and third-nearest neighbor interactions are ($J_{ij} = J_1$, $J_2$ and $J_3$).

Regarding the PSMO the values of $J_1 = +42.0$ (K), $J_2 = +39.0$ (K) and $J_3 = +36.0$ (K) are found from the mean field theory [26]. While the magnetic couplings of LSMO are calculated from the DFT method using the total energies of the ferromagnetic state and the antiferromagnetic state. We found the following values: $J_1 = +100.98$, $J_2 = +14.27$ and $J_3 = -1.45$ K. Positive negative values of the magnetic coupling constant indicate an antiferromagnetic ferromagnetic interaction [27].

The internal energy per site is given by:

$$E = \frac{1}{N} \langle H \rangle$$

The magnetization M being proportional to the sum of the spins, we put, for the sake of simplicity $M = \frac{1}{N} \left\langle \sum_i \sigma_i \right\rangle$. The magnetic contribution to the susceptibility is given by:

$$\chi = \frac{N}{k_B T}[\langle M^2 \rangle - \langle M \rangle^2],$$

where T denotes the absolute temperature and $k_B$ is the Boltzmann's constant.

The magnetic contribution to the magnetic specific heat is given by:

$$C_m = \frac{\beta^2}{N}[\langle E^2 \rangle - \langle E \rangle^2]$$

where $\beta = \frac{1}{k_B T}$.

The magnetic entropy is given by:

$$S(T, h) = \int_0^T \frac{C_m}{T'} dT'$$

The magnetic field-induced entropy change is given by:

$$\Delta S(T, h) = \int_0^h \left(\frac{\partial M}{\partial T}\right)_h dh'.$$

## 2.3  Crystal Structure of Manganite Perovskite

The atoms geometric arrangement in a crystal is called the crystal structure. Furthermore, molecules consist of atoms group and have a molecular structure. Therefore, in a molecular crystal, the molecules of a particular structure are grouped together to form another structure which is also called the crystal structure. In three dimensions, there are seven crystal structures: triclinic, monoclinic, orthorhombic, tetragonal, rhombohedral, hexagonal, and cubic.

The basic properties of a crystal depend on its crystal structure. Consequently, detailed knowledge of the crystal structure is important. Moreover, the structure of atoms affects the properties of the material, for example, perovskite oxide, intermetallic and alloys have very good ductility.

The crystal structure has imperfections, such as point defects (dissolved atoms, vacancies) and dislocations, and these imperfections control many properties of the material [28–30].

The structure file of our material contains 20 atoms (4 Pr, 4 Mn and 12 O) with lattice parameters a = 5.90 (Å), b = 7.72 (Å), c = 5.52 (Å) and $\alpha = \beta = \gamma = 90°$ with the space group Pnma (Fig. 2.1). To mimic the realistic concentration of Sr (x = 0.35) in the PrMnO$_3$ Pnma lattice, (1 *1 *5) supercell is formed containing 100 atoms (20 Pr, 20 Mn and 60 O atoms in Fig. 2.1). A Sr/Pr substitution in the supercell

**Fig. 2.1** The structure of
PrMnO$_3$

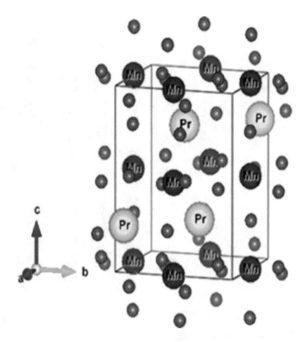

of the total atom of 13 Pr leads to a composition Pr$_{0.65}$Sr$_{0.35}$MnO$_3$ which is even of
the composition Pr$_{0.65}$Sr$_{0.35}$MnO$_3$ prepared experimentally [31].

The structure file of LaMnO$_3$ contains 20 atoms (4 La, 4 Mn and 12 O) with
lattice parameters a = 5.871(Å), b = 7.777(Å), c = 5.586 (Å) and $\alpha = \beta = \gamma = 90°$
with the space group Pnma (Fig. 2.2). To mimic the realistic concentration of Sr (x
= 0.25), we will replace a single atom of La by Sr.

## 2.4  Electronic Properties of Manganite Perovskite

The half-metallic is important for spintronics devices and refrigeration magnetic.
The half-metallic manganite perovskite oxides can be used to achieve 100% spin-
polarized current at the Fermi level. In addition, these materials show excellent
dynamical, thermal, and mechanical properties [32].

Manganese oxide tend to adopt a distorted octahedral molecular structure due
to the Jan-Teller effect (Fig. 2.3). On the other hand, Oxygen vacancy defects play
an important role in forming the properties of manganese-based perovskite oxide
[33]. For example, reported that oxygen vacancies cause significant changes in the
electronic and magnetic structures of La$_{0.66}$Sr$_{0.33}$MnO$_3$ [34]. It was also found that
the introduction of oxygen vacancies causes a shift of the valence band features
toward higher binding energies. The introduction of oxygen vacancies also leads to
an increase of the degree of covalency of Mn bonding.

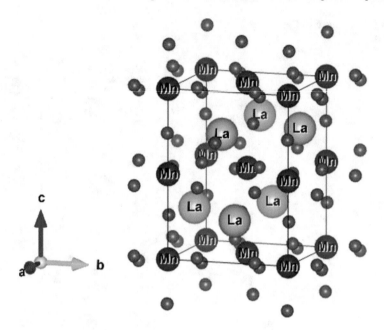

**Fig. 2.2** The structure of LaMnO₃

**Fig. 2.3** Illustration of
MnO₆ octahedron in
manganite

In this part we will study the electronic properties of the LSMO and PSMO such as the total TDOS and partial PDOS density.

In Fig. 2.4, display the total DOS of $Pr_{0.65}Sr_{0.35}MnO_3$ between −7 and 5 eV as a function of photon energy (eV). This figure shows that there is a gap energy close to the Fermi level for spin down, which indicates that this compound has a half metallic character. The sum of the total density of spin up and down is greater than zero, so this compound has a ferromagnetic behavior.

To go deeper to explain the electronic properties, we plotted PDOS Fig. 2.5. The atoms Mn and O have a significant contribution of the total density around the Fermi level $E_F$. While Pr has a strong contribution from energy range −0.53 to 0.85 eV. The contribution from Sr to the total DOS near Fermi level is negligible. Additionally, the Mn atom is described by the 3d orbital, while the O atom is described by the 2p orbital. Therefore, there is hybridization that occurs with Mn-d and O-p [35]. This

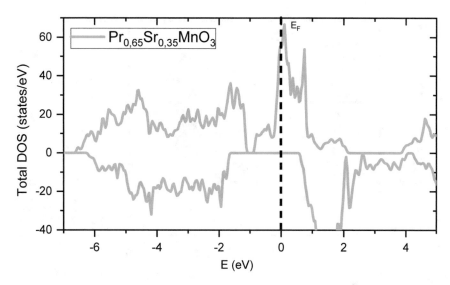

**Fig. 2.4**   The total DOS of $Pr_{0.65}Sr_{0.35}MnO_3$ compounds from FLAPW calculations

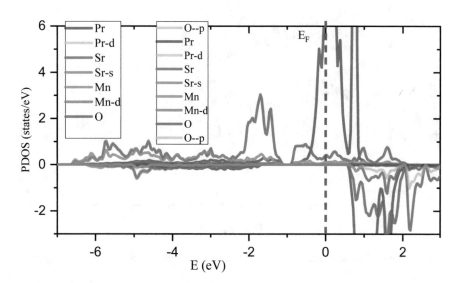

**Fig. 2.5**   The PDOS of $Pr_{0.65}Sr_{0.35}MnO_3$ compounds from FLAPW calculations

suggests that the half-metallicity and the magnetic spin moment are mainly due to the 2p (O)–3d (Mn) coupling.

From the Mn-d orbital projected density of states (PDOS) (Fig. 2.6a), one can clearly see that the $dx^2-y^2$ and dxz orbitals are predominant at $E_F$, while the dxy, dyz and $dz^2$ orbitals are predominant at $[-2.36\,eV, -1.13\,eV]$. From O-p we have strong contribution of px, py and pz spin up in energy range $-6.8$ to $-1.11\,eV$ (Fig. 2.6b).

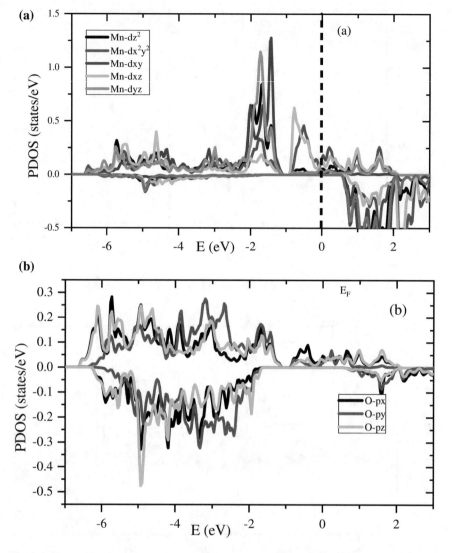

**Fig. 2.6** Mn-d orbital projected partial (**a**) and O-p orbital projected partial (**b**) of $Pr_{0.65}Sr_{0.35}MnO_3$

To calculate the percentage of the spin polarization at the Fermi level $E_F$, we used this equation:

$$p(N) = \frac{N\uparrow - N\downarrow}{N\uparrow + N\downarrow},$$

**Table 2.1** The magnetic moment of Mn, and spin polarization

| Compound | Magnetic moment ($\mu_B$) | SP (%) |
|---|---|---|
| $Pr_{0.65}Sr_{0.35}MnO_3$ | 3.37617 | 100 |

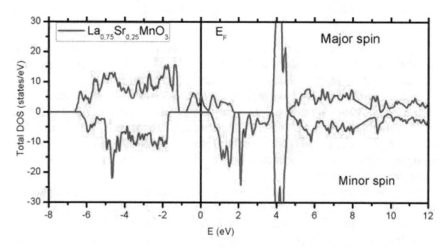

**Fig. 2.7** The total DOS of $La_{0.75}Sr_{0.25}MnO_3$ compounds from FLAPW calculations

where N is the density of states at $E_F$ for spin up ↑ or spin down ↓ [36]. It is found that this system presents 100% spin polarization. The calculated Mn spin magnetic moment is 3.37$\mu_B$ calculated by GGA-PBE (Table 2.1).

The total density (TDOS) as function of photon energy projected between −8 and 2 eV for $La_{0.75}Sr_{0.25}MnO_3$ are illustrated in Fig. 2.7. Here, the Fermi level $E_F$ is taken as reference. As we see, this DOS is not symmetrical with respect to the axis of energy, emphasizing that the magnetic moments carried by the Mn atoms are ferromagnetic. We can also note the absence of the gap at Fermi level for spin up, demonstrates the compound is half-metallic character.

To study the magnetic behavior in more detail, we calculated the exchange energy using ab initio calculations. This exchange energy is defined as $E_{ex} = (E_{AFM} - E_{FM})$, where $E_{AFM}$ and $E_{FM}$ are the energies of the Mn atoms with antiferromagnetic (AFM) and ferromagnetic (FM) coupling. The differences energy between these magnetic configurations $\Delta E_i$ (i = 1–3) is calculated for the three distinct antiferromagnetic configurations A-AFM, C-AFM and G-AFM using the DFT method [37–39]. The obtained value is cited in Table 2.2. The obtained values are positive, which confirms that the ground state FM is more stable than the AFM states [40].

Regarding spin polarization information was obtained by calculating the degree of this parameter in $La_{0.75}Sr_{0.25}MnO_3$ from DOS data around $E_F$. As mentioned above we neglected the spin–orbit coupling and using the simple equation: $p(N) = \frac{N\uparrow - N\downarrow}{N\uparrow + N\downarrow}$. It is found that this system presents 100% spin polarization. The strength of the spin polarization depends on the hybridization effect.

**Table 2.2** Calculated energies of the ferromagnetic state ($E_{FM}$) and antiferromagnetic states ($E_{AFM}$), the energy differences ($\Delta E$) between $E_{AFM}$ and $E_{FM}$ for the LSMO compound

| Compound | $m(\mu_B)$ | $\Delta E_1(R_y)$ | $\Delta E_2(R_y)$ | $\Delta E_3(R_y)$ |
|---|---|---|---|---|
| LSMO | 3.35 | 0.038578 | 0.009217 | 0.052509 |

Figure 2.8 shows the total density of each atom, the Mn and O atoms have a strong contribution of the total density in the vicinity of the Fermi EF level, and a weak contribution of the electron density for the Sr and La atoms.

To fully understand we have plotted the partial densities of the orbitals of the valence band. According to Fig. 2.9 a strong contribution from the O-2p and Mn-3d orbitals, but, the Sr-5s and La-4d orbitals are weakly contributed. Therefore, the Mn atom is dominant by the 3d orbital and the O atom is dominant by the 4p orbital. The strong contribution of the electronic density of Mn-3d is due to the hybridization occurring with O-2p states [35, 41].

It is well-known that the atom Mn in the free state has the electronic configuration of the outer layer $3d^5 4s^2$, where the five orbitals 3d is $x^2-y^2$, $z^2$, xy, xz et yz are occupied by a parallel spin electron. The three 2p orbitals of the oxygen atom are: px, py and pz [42]. In the perovskite manganese $La_{0.75}Sr_{0.25}MnO_3$, the crystal field partially lifts degeneration into two $t_{2g}$ (xy, xz et yz) and e.g. ($x^2-y^2$, $z^2$) sub-levels that occur by hybridization of the 2p oxygen orbit $MnO_6$ (Fig. 2.9) [43, 44].

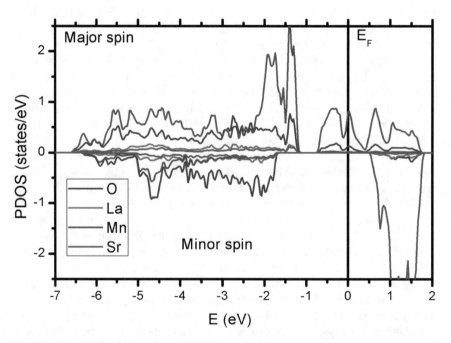

**Fig. 2.8** PDOS of LSMO from FLAPW calculations

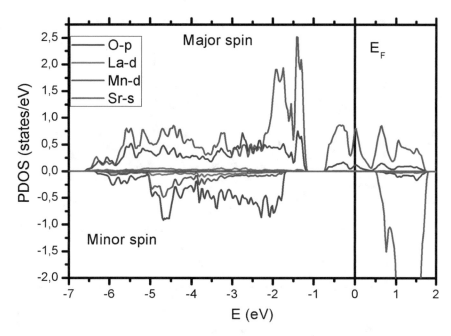

**Fig. 2.9** PDOS for Mn-3d, La-5d, O-4p and Sr-5s of $La_{0.75}Sr_{0.25}MnO_3$

The strong hybridization of Mn-3d and O-2p can also be explained by electronegativity (Fig. 2.10). The role of the hybridization between Mn 3d and O 2p orbitals in the existence of the Griffiths phase in $La_{0.85}Ca_{0.15}MnO_3$ has been investigated [44]. The electronegativity of the elements composing our system is Mn (1.55), O (3.44), La (1.1) and Sr (0.95) (Pauling scale). This suggests that the half-metallicity and the magnetic spin moment are mainly due to the 2p (O)–3d (Mn) coupling, since the highest electronegativity belongs to the O atom [45]. The magnetic spin moment of Mn is 3.35 $\mu_B$. This value is in excellent agreement with experimental values 3.37 $\mu_B$ [46]. So, we have $Mn^{3+}$–$O_2$–$Mn^{4+}$ configuration [47], there will be a ferromagnetic coupling which results from a mixed situation of the two configurations ($Mn^{3+}$–$O_2$–$Mn^{3+}$ and $Mn^{4+}$–$O_2$–$Mn^{4+}$) [48].

## 2.5 Magnetic and Magnetocaloric Properties of Manganite Perovskite

As known, there are four types of magnetic properties: ferromagnetic, ferrimagnetic, antiferromagnetic and paramagnetic. The results obtained using the DFT method showed that the two materials studied have a ferromagnetic behavior. So, these two materials have the possibility of becoming permanent magnets. Therefore, manganese element gives them strong magnetization ability.

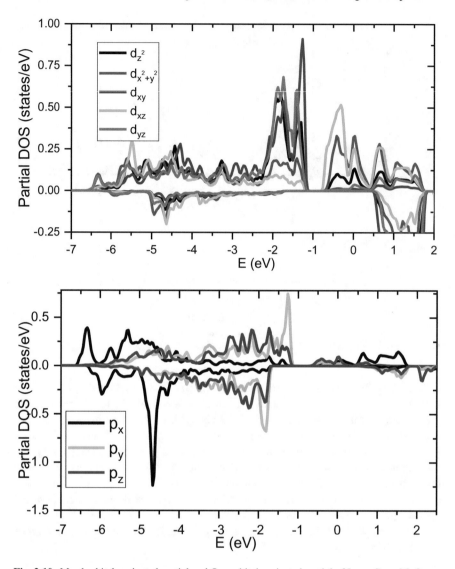

**Fig. 2.10** Mn-d orbital projected partial and O-p orbital projected partial of $La_{0.75}Sr_{0.25}MnO_3$

The magnetocaloric effect (MEC) corresponds to a variation the temperature adiabatic ($\Delta T_{ad}$) or the isothermal variation of magnetic entropy ($\Delta S_M$) of a solid under external magnetic field. We observe a passage through extrema (minima or maxima) at the transition temperatures. When the field is varied from $H_1$ to $H_2$ ($H_2 > H_1$) adiabatically and reversibly, the system undergoes a rise in temperature and a decrease in magnetic entropy $\Delta S_M$. This rise $\Delta T_{ad}$ can be estimated as the isentropic difference of the S(T) curves at $H_1$ and $H_2$ shown in the figure below Fig. 2.11.

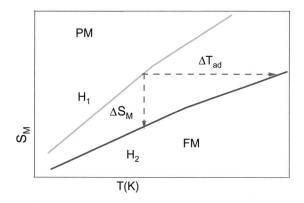

**Fig. 2.11** Thermodynamic principle of the magnetocaloric effect (PM is paramagnetic and FM is ferromagnetic)

The total entropy as a function of temperature for two external applied magnetic fields $H_1 = 0$ and $H_2 > 0$. The inset shows the thermal variation of the MEC typical of a ferromagnetic material (the quantities $\Delta S_M$ and $\Delta T_{ad}$ is indicated by arrows). The isothermal variation of the magnetic entropy $\Delta S_M$ is obtained by taking the difference between the entropy values of the final state and that of the initial state of any isothermal process. The variation of the adiabatic temperature $\Delta T_{ad}$ and the variation of the magnetic entropy $\Delta S_M$ therefore represent the two physical quantities making it possible to quantify the magnetocaloric effect.

This effect is the basis of the new technology of magnetic refrigerators. On the other hand, the magnetocaloric effect was discovered in 1881 by Warburg [49]. He observed around the Curie temperature of iron ($T_C \approx 1043$ K) a rise (lower) in temperature during a sudden application (or disapplication) of an external magnetic field. In 1918, Weiss and Piccard theoretically expressed this phenomenon and assigned the name magnetocaloric effect. Then in 1926 Debye then Giauque [50, 51] explained this phenomenon thermodynamically and suggested its use in processes allowing to reach low temperatures by a process called adiabatic demagnetization.

The magnetocaloric properties of rare earth manganite were unveiled [52]. In the following we will cite some studies of MEC in doped manganates $T_{1-x}Sr_xMnO_3$ with (T = La, Pr) and (x = 0.35 and 0.25) using Monte Carlo simulations:

We have illustrated in Fig. 2.12 the thermal magnetization as function of temperature. From this figure, we deduce that the Curie temperature is equal to $T_C = 294$ K. This value is in good agreement with that found experimentally from Ref. [31] (see Table 2.3). The magnetization curve shows a classical phase transition of the second order at $T_C$ (ferromagnetic/paramagnetic).

Figure 2.13, display the variation of thermal magnetic entropy change as function of temperature under external magnetic field of $Pr_{0.65}Sr_{0.35}MnO_3$. The value of $\Delta S^{max}$ increases with increasing the values of external magnetic field H. While the Curie temperature remains stable. The maximum of magnetic entropy is $-\Delta S^{max} = 1.98$ J K$^{-1}$ kg$^{-1}$. This value is in good agreement with the experimental one in Ref. [31] (see Table 2.3).

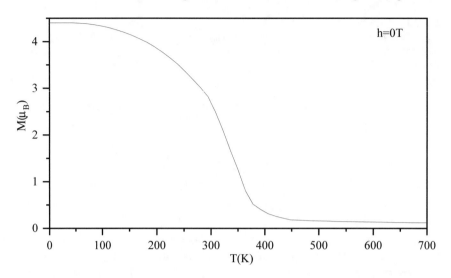

**Fig. 2.12** The thermal magnetization for h = 0 T

**Table 2.3** The values of $T_C$, $\Delta S^{max}$, $\Delta T_{ad}$ and $C_P$ obtained by Ref. [31] and Monte Carlo simulations

| Perovskite | $T_C$ (K) | $|\Delta S^{max}|$ (J/kg K) at h = 1 T | $\Delta T_{ad}$ (K) at h = 1T | $C_P$ (J/(kg K)) at $T_C$ |
|---|---|---|---|---|
| $Pr_{0.65}Sr_{0.35}MnO_3$ | 295 [31] | 2.3 [31] | 1.1 [31] | 580 [31] |
| | 294 Monte Carlo simulations (MCS) | 2 MCS | 1.13 MCS | 581 MCS |

The thermal adiabatic temperature change $\Delta T_{ad}$ and heat specific $C_P$ are shown in Fig. 2.14 for h = 1 T. The maximum of adiabatic temperature change (Fig. 2.14a) and heat specific (Fig. 2.14b) are situated at the transition temperature $T_C = 294$ K. This value of $T_C$ is confirmed by the magnetization curve (Fig. 2.12) and magnetic entropy change curve (Fig. 2.13) which exhibits a clear transition around 294 K. The values obtained of $\Delta T_{ad}$ and $C_P$ are comparable with that given by experimental results of Ref. [31] (see Table 2.3).

We show in Fig. 2.15, the variation of relative cooling power with the magnetic field for $Pr_{0.65}Sr_{0.35}MnO_3$. RCP varies linearly with magnetic field h. The maximum value of RCP is 17557 J/kg is given for h = 5T.

These results that we obtained are reasonable because there is always a giant magnetocaloric effect and a large maximum of the RCP for rare-earth compounds containing large magnetic moments [53]. Figure 2.16 display the thermal dependence of relative cooling power for $Pr_{0.65}Sr_{0.35}MnO_3$ for a several magnetic fields. The maximum RCP values increase with the increase the external magnetic field. While

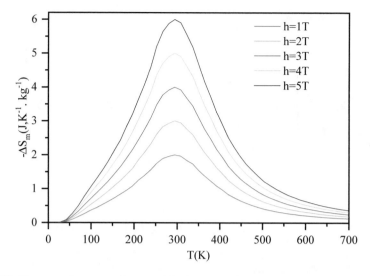

**Fig. 2.13** The magnetic entropy change $-\Delta S_m$ for a several magnetic fields

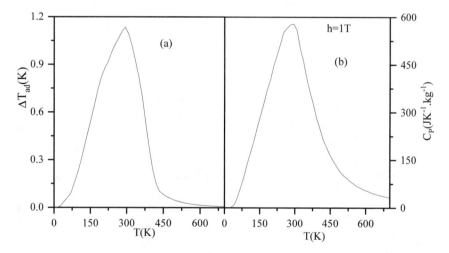

**Fig. 2.14** The adiabatic temperature change $\Delta T_{ad}$ (**a**) and heat specific $C_P$ (**b**) for h = 1 T

the critical temperature $T_C$ remains constant. These RCP values increase until reached their saturation for each value of magnetic field.

Figure 2.17a shows the variation of the magnetization versus the temperature. The transition from a ferromagnetic state (FM) to a paramagnetic state (PM) is observed at near the Curie temperature $T_C$.

The obtained value of $T_C$ is 310 K. This value of the temperature $T_C$ is determined from the curve dM/dT as shown in Fig. 2.17b. This value is very close to the experimental value 313 K [54].

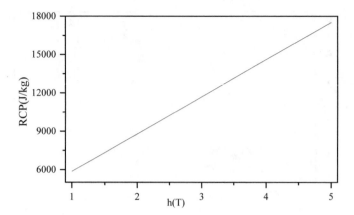

**Fig. 2.15**  The field dependence of relative cooling power for $Pr_{0.65}Sr_{0.35}MnO_3$

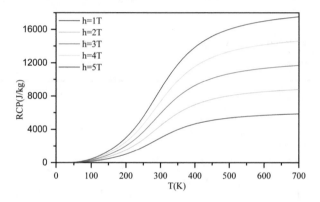

**Fig. 2.16**  The thermal dependence of relative cooling power for a several magnetic fields

**Fig. 2.17  a** Magnetization as a function of temperatures with different external magnetic fields H = 0 T, 4 T and 6 T. **b** dM/dT as a function of temperatures with different external magnetic fields H = 0 T, 4 T and 6 T

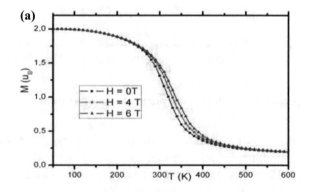

**Fig. 2.17** (continued)

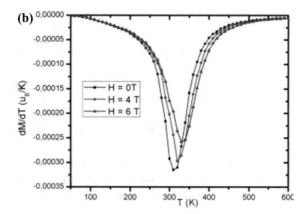

To understand the magnetic behavior of the compound $La_{0.75}Sr_{0.25}MnO_3$ in the paramagnetic region (above $T_C$), we have shown in Fig. 2.18 the magnetic susceptibility $\chi$ depending on the temperature under the influence of the magnetic field of 0–6 T. The maximum of the susceptibility is strongly reduced, and it is smeared out over a broader temperature interval. We also observe a shift of their maximum to higher temperatures at about the same rate as the shift of $T_C$.

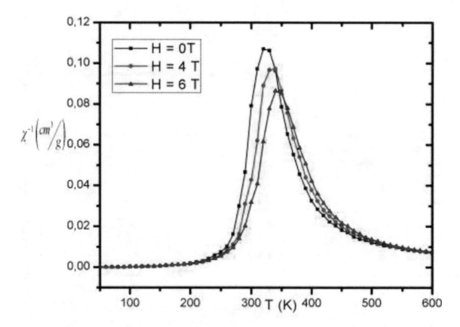

**Fig. 2.18** Susceptibility versus temperatures with different external magnetic fields H = 0 , 4 and 6 T

Figure 2.19 illustrates the variation of specific heat as a function of temperature of $La_{0.75}Sr_{0.25}MnO_3$ for several magnetic fields. The increase of the magnetic field leads to the displacement of the maximum which corresponds to the Curie temperature monotonously towards higher temperatures. On the other hand, the influence of the external field leads to the reduction and spread of the maximum of the specific heat over a wider temperature range.

Figure 2.20 shows the reciprocal change in magnetic entropy of the $La_{0.75}Sr_{0.25}MnO_3$ as a function of temperature in various magnetic fields applied. As expected, a small displacement of the maximum MCE to higher temperatures at about the same rate as the shift of $T_C$ and the effect is smeared out over a broader temperature interval. The maximum value of magnetocaloric effect is found to increase monotonically with the applied magnetic field increasing and reaches a value of 9.23 J/K kg around $T_C$ for a magnetic field change from 0 to 6 T.

**Fig. 2.19** Specific heat of LSMO under different magnetic fields. H = 0 , 4 and 6 T

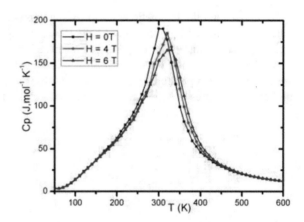

**Fig. 2.20** Temperature dependence of the magnetic entropy change for different external magnetic fields H = 4 and 6 T by Monte Carlo simulations

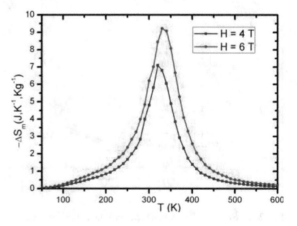

**Table 2.4** Summary of the magnetocaloric properties obtained by the Monte Carlo method of $La_{0.75}Sr_{0.25}MnO_3$ compared to other magnetic materials

| Perovskite | $\mu_0 H(T)$ | $T_C$ (K) | $|\Delta S^{max}|$ (J/kg. K) | References |
|---|---|---|---|---|
| $La_{0.75}Sr_{0.25}MnO_3$ | 6 | 310 | 9.23 | This work |
| $La_{0.75}Sr_{0.25}MnO_3$ | 4 | 310 | 7 | This work |
| $La_{0.75}Sr_{0.25}MnO_3$ | 2 | 332 | 2 | [19] |
| $La_{0.75}Sr_{0.25}MnO_3$ | 1.5 | 340 | 1.5 | [55] |
| $La_{0.8}K_{0.2}MnO_3$ | 5 | 281 | 3.71 | [56] |
| $La_{0.67}Ba_{0.33}MnO_3$ | 2 | 350 | 1.72 | [24] |
| $La_{0.67}Ba_{0.33}MnO_3$ | 5 | 292 | 1.48 | [57] |
| $La_{0.67}Sr_{0.33}MnO_3$ | 2 | 377 | 2.02 | [58] |
| $La_{0.67}Sr_{0.33}MnO_3$ | 2 | 354 | 1.15 | [59] |

To evaluate the applicability of our compound as magnetic refrigerant, the obtained values of magnetic entropy in our study, compared to other magnetic materials, are summarized in Table 2.4.

## 2.6 Conclusions

In this chapter, we studied the magnetic, electronic and magnetocaloric properties of $La_{0.75}Sr_{0.25}MnO_3$ and of $Pr_{0.65}Sr_{0.35}MnO_3$ perovskites. We used the DFT method to determine the electronic and magnetic properties. We used the Monte Carlo method to determine the magnetocaloric properties of both compounds. Both compounds crystallize in the orthorhombic structure, space group Pnma. The spin polarization for two compounds shows a half metallic character with the ferromagnetic coupling of the Mn spin. The magnetic properties of these compounds exhibit a single transition from the FM state to the PM state. The values obtained from $T_C$ are 294 and 310 K for PSMO and LSMO respectively. These results are comparable to those obtained experimentally. The results obtained from magnetic entropy and specific heat are processed and analyzed. We found results very consistent with those found experimentally. The maximum magnetic entropy is at the Curie temperature. Its increases with increasing magnetic field. The adiabatic temperature and the specific heat are treated. The maximum of the curves is at Curie temperature. The variation of the RCP values as a function of the values of the magnetic fields and the temperature is found. RCP values increase with increasing magnetic field.

# References

1. P.V. Balachandran, A.A. Emery, J.E. Gubernatis T. Lookman, C. Wolverton, A. Zunger, Predictions of new $ABO_3$ perovskite compounds by combining machine learning and density functional theory. Phys. Rev. Mater. **2**(4), 043802 (2018)
2. F. Neese, Prediction and interpretation of the 57Fe isomer shift in Mössbauer spectra by density functional theory. Inorg. Chim. Acta. Chim. Acta **337**, 181–192 (2002)
3. M. Barhoumi, The density functional theory and beyond: example and applications, in *Density Functional Theory-Recent Advances, New Perspectives and Applications* (IntechOpen, 2021)
4. P. Kratzer, J. Neugebauer, The basics of electronic structure theory for periodic systems. Front. Chem. **7**, 106 (2019)
5. E.E. Carpenter, U.S. Patent Application No. 16/768,173. (2020)
6. A. Barman, S. Kar-Narayan, D. Mukherjee, Caloric effects in perovskite oxides. Adv. Mater. Interfaces **6**(15), 1900291 (2019)
7. R.V. Demin, L.I. Koroleva, Influence of a magnetic two-phase state on the magnetocaloric effect in the $La_{1-x}Sr_xMnO_3$ manganites. Phys. Solid State **46**(6), 1081–1083 (2004)
8. A. Szewczyk, H. Szymczak, A. Wiśniewski, K. Piotrowski, R. Kartaszyński, B. Dabrowski, Z. Bukowski, Magnetocaloric effect in $La_{1-x}Sr_xMnO_3$ for x = 0.13 and 0.16. Appl. Phys. Lett. **77**(7), 1026–1028 (2000)
9. T. Geng, N. Zhang, Electronic structure of the perovskite oxides $La_{1-x}Sr_xMnO_3$. Phys. Lett. A **351**(4–5), 314–318 (2006)
10. B. Arun, M. Athira, V.R. Akshay, B. Sudakshina, G.R. Mutta, M. Vasundhara, Investigation on the structural, magnetic and magnetocaloric properties of nanocrystalline Pr-deficient $Pr_{1-x}Sr_xMnO_{3-\delta}$ manganites. J. Magn. Magn. Mater.Magn. Magn. Mater. **448**, 322–331 (2018)
11. T.D. Thanh, T.A. Ho, T.V. Manh, T.L. Phan, S.C. Yu, Large magnetocaloric effect around room temperature in double-exchange ferromagnets $Pr_{1-x}Sr_xMnO_3$ with short-range interactions. IEEE Trans. Magn.Magn. **50**(11), 1–4 (2014)
12. T.A. Ho, T.D. Thanh, Y. Yu, D.M. Tartakovsky, T.O. Ho, P.D. Thang, S.C. Yu, Critical behavior and magnetocaloric effect of $Pr_{1-x}Ca_xMnO_3$. J. Appl. Phys. **117**(17), 17D122 (2015)
13. Y. Yu, D.T. Pham, A.T. Le, Critical behavior and magnetocaloric effect of $Pr_{1x}Ca_xMnO_3$. J. Appl. Phys. **117**(17D122) (2015)
14. H. Terashita, J.J. Garbe, J.J. Neumeier, Compositional dependence of the magnetocaloric effect in $La_{1-x}Ca_xMnO_3$ ($0 \le x \le 0.52$). Phys. Rev. B **70**(9), 094403 (2004)
15. B.R. Dahal, K. Schroeder, M.M. Allyn, R.J. Tackett, Y. Huh, P. Kharel, Near-room-temperature magnetocaloric properties of $La_{1-x}Sr_xMnO_3$ (x = 0.11, 0.17, and 0.19) nanoparticles. Mater. Res. Expr. **5**(10), 106103 (2018)
16. A.R. Dinesen, Magnetocaloric and magnetoresistive properties of $La_{0.67}Ca_{0.33-x}Sr_xMnO_3$. Disertasi, Technical University of Denmark, 2004
17. K.P. Shinde, N.G. Deshpande, T. Eom, Y.P. Lee, S.H. Pawar, Solution-combustion synthesis of $La_{0.65}Sr_{0.35}MnO_3$ and the magnetocaloric properties. Mater. Sci. Eng. B **167**(3), 202–205 (2010)
18. J. Jiang, Q.M. Chen, X. Liu, First-principles study on the electronic structure and optical properties of $La_{0.75}Sr_{0.25}MnO_{3-\sigma}$ materials with oxygen vacancies defects. Curr. Appl. Phys. **18**(2), 200–208 (2018)
19. M. Pekala, K. Pekala, V. Drozd, J.F. Fagnard, P. Vanderbemden, Magnetocaloric effect in $La_{0.75}Sr_{0.25}MnO_3$ manganite. J. Magn. Magn. Mater. **322**(21) (2010)
20. S. Mollah, Thermal hysteresis in resistivity and magnetization of PrCa(Sr)MnO. Mod. Phys. Lett. B **22**(32), 3241–3248 (2008)
21. A.K. Mishra, A.J. Darbandi, P.M. Leufke, R. Kruk, H. Hahn, Room temperature reversible tuning of magnetism of electrolyte-gated $La_{0.75}Sr_{0.25}MnO_3$ nanoparticles. J. Appl. Phys. **113**(3), 033913 (2013)

22. B. Kohler, S. Wilke, M. Scheffler, R. Kouba, C. Ambrosch-Draxl, Force calculation and atomic-structure optimization for the full-potential linearized augmented plane-wave code WIEN. Comput. Phys. Commun.. Phys. Commun. **94**(1), 31–48 (1996)

23. H.Y.S. Shalash, FP-LAPW study of phase Changesin An (A= Al, IN, and B) under high pressure. Doctoral dissertation, 2009

24. K. Schwarz, DFT calculations of solids with LAPW and WIEN2k. J. Solid-State Chem. **176**(2), 319–328 (2003)

25. J.P. Bergsma, Combining mean field calculations with Monte Carlo simulations for polymer gels and dendrimers. Doctoral dissertation, Wageningen University and Research, 2019

26. W.E. Holland, H.A. Brown, Application of the Weiss molecular field theory to the B-site spinel. Phys. Status Solidi (A) **10**(1), 249–253 (1972)

27. G. Kadim, R. Masrour, A. Jabar, A comparative study between GGA, WC-GGA, TB-mBJ and GGA+U approximations on magnetocaloric effect, electronic, optic and magnetic properties of $BaMnS_2$ compound: DFT calculations and Monte Carlo simulations. Phys. Scr. **96**(4), 045804 (2021)

28. E.J. Mittemeijer, The crystal imperfection; lattice defects, in *Fundamentals of Materials Science* (Springer, Berlin, Heidelberg, 2010), pp. 201–244

29. M. Razeghi, Defects, in *Fundamentals of Solid-State Engineering* (Springer, Boston, MA, 2009), pp. 1–21

30. P. Kofstad, Defects and transport properties of metal oxides. Oxid. Met. **44**(1), 3–27 (1995)

31. F. Guillou, U. Legait, A. Kedous-Lebouc, V. Hardy, Development of a new magnetocaloric material used in a magnetic refrigeration device, in *EPJ Web of Conferences*, vol. 29 (EDP Sciences, 2012), p. 00021

32. S.A. Khandy, D.C. Gupta, Investigation of the transport, structural and mechanical properties of half-metallic $REMnO_3$ (RE = Ce and Pr) ferromagnets. RSC Adv. **6**(100), 97641–97649 (2016)

33. B. Udeshi, H. Boricha, B. Rajyaguru, K. Gadani, K.N. Rathod, D. Dhruv, N.A. Shah, Electrical behavior and structure–property correlations in $La_{1-x}Pr_xMnO_3$ ($0\leq x\leq 1$) ceramics. Ceram. Int. **45**(1), 1098–1109 (2019)

34. S. Picozzi, C. Ma, Z. Yang, R. Bertacco, M. Cantoni, A. Cattoni, F. Ciccacci, Oxygen vacancies and induced changes in the electronic and magnetic structures of $La_{0.66}Sr_{0.33}MnO_3$: a combined ab initio and photoemission study. Phys. Rev. B **75**(9), 094418 (2007)

35. M. Chakraborty, P. Pal, B.R. Sekhar, Half metallicity in $Pr_{0.75}Sr_{0.25}MnO_3$: a first principle study. Solid State Commun. **145**(4), 197–200 (2008)

36. I. Galanakis, P. Mavropoulos, Spin-polarization and electronic properties of half-metallic Heusler alloys calculated from first principles. J. Phys. Condens. MatterCondens. Matter **19**(31), 315213 (2007)

37. Z. Wu, J. Yu, S. Yuan, Strain-tunable magnetic and electronic properties of monolayer CrI3. Phys. Chem. Chem. Phys. (2019)

38. S. Naji, A. Benyoussef, A. El Kenz, H. Ez-Zahraouy, M. Loulidi, Monte Carlo study of phase transitions and magnetic properties of $LaMnO_3$: Heisenberg model. Phys. A **391**(15), 3885–3894 (2012)

39. G. Kadim, R. Masrour, A. Jabar, Large magnetocaloric effect, magnetic and electronic properties in $Ho_3Pd_2$ compound: Ab initio calculations and Monte Carlo simulations. J. Magn. Magn. Mater.Magn. Magn. Mater. **499**, 166263 (2020)

40. J.E. Medvedeva, V.I. Anisimov, O.N. Mryasov, A.J. Freeman, The role of Coulomb correlation in magnetic and transport properties of doped manganites: $La_{0.5}Sr_{0.5}MnO_3$ and $LaSr_2Mn_2O_7$. J. Phys. Condens. Matter **14**(17), 4533 (2002)

41. I.V. Alabugin, S. Bresch, M. Manoharan, Hybridization trends for main group elements and expanding the Bent's rule beyond carbon: more than electronegativity. J. Phys. Chem. A **118**(20), 3663–3677 (2014)

42. B.R.K. Nanda, S. Satpathy, Effects of strain on orbital ordering and magnetism at perovskite oxide interfaces: $LaMnO_3/SrMnO_3$. Phys. Rev. B **78**(5), 054427 (2008)

43. Z. Liao, N. Gauquelin, R.J. Green, S. Macke, J. Gonnissen, S. Thomas, P. Hansmann, Thickness dependent properties in oxide heterostructures driven by structurally induced metal-oxygen hybridization variations. Adv. Funct. Mater.Funct. Mater. **27**(17), 1606717 (2017)
44. H.G. Zhang, J.J. Shi, Y.T. Li, X.G. Dong, X.P. Ge, H. Liu, Y.K. Tang, The role of the hybridization between Mn-3d and O-2p orbitals in the existence of the Griffiths phase in $La_{0.85}Ca_{0.15}MnO_3$. J. Phys. Condens. Matter **26**(14), 145601 (2014)
45. J. He, P. Zhou, N. Jiao, S.Y. Ma, K.W. Zhang, R.Z. Wang, L.Z. Sun, Magnetic exchange coupling and anisotropy of 3d transition metal nanowires on graphyne. Sci. Rep. **4**, 4014 (2014)
46. R. Rahmani, B. Amrani, K.D. Khodja, A. Boukhachem, P. Aubert, Systematic study of elastic, electronic, and magnetic properties of lanthanum cobaltite oxide. J. Comput. Electron.Comput. Electron. **17**(3), 920–925 (2018)
47. J. He, A. Borisevich, S.V. Kalinin, S.J. Pennycook, S.T. Pantelides, Control of the structural and magnetic properties of perovskite oxide ultrathin films through the substrate symmetry effect. Phys. Rev. Lett. **105**, 227203 (2010)
48. C. Zener, Interaction between the d-shells in the transition metals. II. Ferromagnetic compounds of manganese with perovskite structure. Phys. Rev. **82**(3), 403 (1951)
49. E. Warburg, Magnetische untersuchungen. Ann. Phys. **249**(5), 141–164 (1881)
50. P. Debye, Einige bemerkungen zur magnetisierung bei tiefer temperatur. Ann. Phys. **386**(25), 1154–1160 (1926)
51. W.F. Giauque, A thermodynamic treatment of certain magnetic effects. A proposed method of producing temperatures considerably below 1 absolute. J. Am. Chem. Soc. **49**(8), 1864–1870 (1927)
52. D.T. Morelli, A.M. Mance, J.V. Mantese, A.L. Micheli, Magnetocaloric properties of doped lanthanum manganite films. J. Appl. Phys. **79**(1), 373–375 (1996)
53. A.M. Tishin, Y.I. Spichkin, *The Magnetocaloric Effect and Its Applications* (CRC Press, 2016)
54. L.Q. Wu, W.H. Qi, X.S. Ge, D.H. Ji, Z.Z. Li, G.D. Tang, W. Zhong, Study of the dependence of the magnetic moment of $La_{1-x}Sr_xMnO_3$ on the Sr doping level x. Europhys. Lett.. Lett. **120**(2), 27001 (2018)
55. Z.B. Guo, W. Yang, Y.T. Shen, Y.W. Du, Magnetic entropy change in $La_{0.75}Ca_{0.25-x}Sr_xMnO_3$ perovskites. J. Solid State Commun. **105**, 89–92 (1998)
56. M. Pekala, K. Pekala, V. Drozd, K. Staszkiewicz, J.F. Fagnard, P. Vanderbemden, J. Appl. Phys. **112**, 023906 (2012)
57. D.T. Morelli, A.M. Mane, J.V. Mantese, A.L. Micheli, J. Appl. Phys. **79**, 373 (1996)
58. Y. Xu, M. Meier, P. Das, M.R. Koblischka, U. Hartmann, Perovskite manganites: potential materials for magnetic cooling at or near room temperature. J. Crystal Eng. **5**(2002), 383 (2002)
59. G.F. Wang, L.R. Li, Z.R. Zhao, X.Q. Yu, X.F. Zhang, Structural and magnetocaloric effect of $Ln_{0.67}Sr_{0.33}MnO_3$(Ln = La, Pr and Nd) nanoparticles. J. Ceram. Int. **40**, 16449–16454 (2014)

# Chapter 3
# Study of Magnetocaloric Effect, Electronic and Magnetic Properties of Perovskite Ferrites

**Abstract** In our research, we employed the linear augmented plane wave method based on density functional theory (DFT) in conjunction with Monte Carlo simulations to investigate the magnetocaloric, electronic, and magnetic properties of the perovskite oxide $Ba_{1-x}Sr_xFeO_3$, where x ranged from 0 to 0.2. Our study incorporated various computational approaches, including spin polarization within the generalized gradient approximation (GGA), Hubbard approximation (GGA + U), and the modified Becke-Johnson potential (TB-mBJ). This study provides a comprehensive understanding of the magnetocaloric, electronic, and magnetic behaviors in Ba1-xSrxFeO3, shedding light on the impact of Sr substitution and offering valuable insights for potential applications.

## 3.1 Introduction

Fe-based perovskite oxides with general chemical formula $AFeO_3$ are a kind of very important functional material [1]. Since the crystal structures and A-Fe ionic groups are quite flexible. On the other hand, these compounds have very important physical properties. Therefore, there is a unique charge transition of $Fe^{4+}$ ions ($2Fe^{4+} \rightarrow Fe^{3+} + Fe^{5+}$) as well as a particular helical spin structure [2]. These important properties can be modified to suit desired applications. This is done by doping of non-magnetic elements in $AFeO_3$ because it's the only means to control the physical properties of this perovskite, and thus the possibility of producing multifunctional perovskite oxide compounds that are suitable for the desired applications [3–6].

Our study will be oriented on $BaFeO_3$ and $Ba_{0.8}Sr_{0.2}FeO_3$ which has been studied theoretically and experimentally to reveal the magnetic, magnetocaloric, dielectric and optical properties and the exchange mechanisms between its atoms in the 3 states: bulk, thin film, or nanoparticles [7–9]. The Sr substitute Ba plays an important role in controlling the magnetocaloric effect, high ionic and electronic conductivities and chemical stability in reducing atmospheres. In addition to that, the control of the relatively low coefficients of thermal expansion and in achieving good chemical stability [10–12].

© The Author(s), under exclusive license to Springer Nature Switzerland AG 2024
R. Masrour, *Magnetoelectronic, Optical, and Thermoelectric Properties of Perovskite Materials*, SpringerBriefs in Materials,
https://doi.org/10.1007/978-3-031-48967-9_3

Theoretical explanation for the electronic and magnetic properties and magnetic effect of $Ba_{1-x}Sr_xFeO_3$ compound with x = 0 and 0.2 has not been explored yet. For this reason, we tried in this chapter to explain the electronic and magnetic properties using the density functional theory DFT and DFT + U technique. Our result has been processed using the GGA and GGA + U approximations as well as the mBJ approximation to conduct a comparative study between the three calculations. While we studied the effect of magnetism using Monte Carlo simulation. Where the exchange coupling constants and magnetic moments are calculated from the DFT and DFT + U technique. We will apply the Hubbard U parameter to obtain an accurate description of the localized Fe-3d states.

## 3.2  Density-Functional Theory and Monte Carlo Simulations

Our calculations were carried out using DFT and DFI + U calculations as simulated in the Wien2k computational code [13]. The electron exchange–correlation energy was treated within the generalized gradient approximation (GGA), GGA + U and TB-mBJ [14]. The TB-mBJ approximation produces gap and magnetic moment values precise. GGA + U Approximation adds some exchange to orbitals of specific angular character (d or f) [15, 16]. In our work we use the TB-mBJ and GGA + U in order to obtain an accurate description of the localized Fe-3d states [17].

The Modified Becke–Johnson (mBJ) proposed by Tran and Blaha has the following form:

$$U_{\chi,\sigma}^{mBJ}(r) = cU_{\chi,\sigma}^{BR}(r) + (3c - 2)\frac{1}{\pi}\sqrt{\frac{5}{12}}\sqrt{\frac{2t_\sigma(r)}{\rho_\sigma(r)}}.$$

Or $\rho(r) = \sum_{i=1}^{n_\sigma}|\psi_{i,\sigma}(r)|^2$ is the density of electrons, $t_\sigma(r) = \frac{1}{2}\sum_{i=1}^{n_\sigma}\nabla\psi_{i,\sigma}^*(r)\nabla\psi_{i,\sigma}(r)$ is the density of kinetic energy and $U_{\chi,\sigma}^{BR}(r) = -\frac{1}{b_\sigma(r)}\left(1 - e^{-x_\sigma(r)} - \frac{1}{2}x_\sigma(r)e^{-x_\sigma(r)}\right)$ is the potential of Becke-Roussel (BR) [18] that has been proposed to model the Coulomb potential created by the exchange hole. The index $\sigma$ is the spin notation. An important issue with the GGA + U approach is the choice of the parameter U. The number of different values of U ranging from 4 to 6 eV has been suggested [19–21]. The first step in this kind of calculation consists in specifying the values of the important parameters, which influence the time and the precision of the calculation: The muffin-tin (MT) radii of Ba, Sr, Fe and O were chosen to be 2.5, 2.5, 1.95 and 1.68 respectively in $Ba_{0.8}Sr_{0.2}FeO_3$ while in $BaFeO_3$ the muffin-tin (MT) radii of Ba, Fe and O were chosen 2.5, 2.4, and 1.75. The cutoff energy is −6.0 Ry. In Fourier series in the interstitial region with a cutoff (cutoff radius) $R_{MTmin}.K_{MAX} = 7$ and $l_{max} = 10$. The wave functions, inside the spheres are expanded up to the calculation gives 300 K-points corresponding to a k-mesh

(6*6*1) for $Ba_{0.8}Sr_{0.2}FeO_3$ and (6 * 6 * 6) for $BaFeO_3$ structure are taken in the irreducible wedge of the Brillouin zone.

Monte Carlo simulations (MCS) an accurate method for studying magnetic and magnetic properties. It is based on the Metropolis algorithm that allows to obtain flexible configurations of Boltzmann statistics [22, 23]. The normalized Boltzmann factor describing the statistical weight at which the formation x appears at thermal equilibrium is:

$$P = \frac{1}{Z} \exp\left(\frac{-H'}{k_B T}\right)$$

where Z is the partition function and H' is Hamiltonian of system. In statistical physics, Several models describe the Hamiltonian of systems. But the Ising model is the most widely used to describe the energy of a system due to its simplicity and richness.

The Hamiltonians of the compounds $BaFeO_3$ and $Ba_{0.8}Sr_{0.2}FeO_3$ which are represented in Fig. 3.1. are written in the following form [24].

$$H' = -\sum_{<i,j>} J_{ij} S_i S_j - H \sum_i S_i$$

where $S_i$ and $S_j$ respectively indicate the spin at the lattice site i and site j. The spin moment of Fe is S = 2 for x = 0 and S = 3/2 for x = 0.2.

The exchange couplings $J_{ij}$ are calculated from the exchange energy (Tables 3.1 and 3.2).

**Fig. 3.1** Structure of $Ba_{1-x}Sr_xFeO_3$ with x = 0 and 0.2

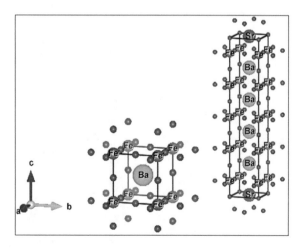

3 Study of Magnetocaloric Effect, Electronic and Magnetic Properties …

**Table 3.1** The difference energy ($\Delta E$) between $E_{AFM}$ and $E_{FM}$ for the $Ba_{1-x}Sr_xFeO_3$ with (x = 0, 0.2) using GGA and GGA + U approximations

| The difference energy | $BaFeO_3$ | | | | $Ba_{0.8}Sr_{0.2}FeO_3$ | | | |
|---|---|---|---|---|---|---|---|---|
| | Theoretical GGA + U | | | Theoretical GGA | Theoretical GGA + U | | | Theoretical GGA |
| | U = 4 eV | U = 5 eV | U = 6 eV | | U = 4 eV | U = 5 eV | U = 6 eV | |
| $\Delta E$ (Ry) | 0.033832 | 0.0341534 | 0.038039 | 0.023496 | 0.029231 | 0.0314996 | 0.032371 | 0.018704 |

**Table 3.2** The exchange coupling parameters of $Ba_{1-x}Sr_xFeO_3$ with (x = 0, 0.2)

| Compounds | $J_1$ (meV) | $J_2$ (meV) | $J_3$ (meV) |
|---|---|---|---|
| $BaFeO_3$ | 0.47 | 0.38 | −0.23 |
| $Ba_{0.8}Sr_{0.2}FeO_3$ | 0.45 | 0.33 | −0.25 |

All magnetic properties depend on the internal energy $E = \frac{1}{N}\langle H \rangle$ and magnetization M $M = \frac{1}{N}\left\langle \sum_i \sigma_i \right\rangle$. These properties are detailed in [24].

## 3.3 Crystal Structure of Ferrite Perovskite

$BaFeO_3$ crystallizes in the cubic P$\overline{m}$3m space group [12]. Lattice parameters a = b = c = 4.012 (Å) and $\alpha = \beta = \gamma = 90°$.

Figure 3.2a shows that Ba is bonded to twelve equivalent O atoms to form $BaO_{12}$ cuboctahedra that share corners with twelve equivalent $BaO_{12}$ cuboctahedra, faces with six equivalent $BaO_{12}$ cuboctahedra, and faces with eight equivalent $FeO_6$ octahedra. All Ba–O bond lengths are 2.85 Å. While Fig. 3.2b shows that Fe is bonded to six equivalent O atoms to form $FeO_6$ octahedra that share corners with six equivalent $FeO_6$ octahedra and faces with eight equivalent $BaO_{12}$ cuboctahedra. All Fe–O bond lengths are 2.02 Å.

To make the realistic concentration of Sr (x = 0.2) in the cubic lattice of $BaFeO_3$, a supercell (1 × 1 × 5) containing 25 atoms (5 Ba, 5 Fe and 15 O atoms in Fig. 3.1) was formed. A Sr / Ba substitution in the supercell of the total atom of 5 Ba leads to a composition $Ba_{0.8}Sr_{0.2}FeO_3$, which is even of the composition $Ba_{0.8}Sr_{0.2}FeO_3$ prepared experimentally. The lattice parameters of $Ba_{0.8}Sr_{0.2}FeO_3$ is a = b = 4.012 Å, c = 20.088 Å and $\alpha = \beta = \gamma = 90°$.

Figure 3.3a shows that there are two non-equivalent Ba sites: The first site, Ba is bonded to twelve O atoms to form $BaO_{12}$ cuboctahedra which share corners with four equivalent $SrO_{12}$ cuboctahedra, corners with eight $BaO_{12}$ cuboctahedra, a face with a $SrO_{12}$ cuboctahedron, faces with five $BaO_{12}$ cuboctahedra and faces with eight $FeO_6$ octahedra. The second Ba site is bonded to twelve O atoms to form $BaO_{12}$ cuboctahedra that share corners with twelve $BaO_{12}$ cuboctahedra, faces with six $BaO_{12}$ cuboctahedra, and faces with eight $FeO_6$ octahedra. Sr is bonded to twelve O atoms to form $SrO_{12}$ cuboctahedra which share corners with four equivalent $SrO_{12}$ cuboctahedra, corners with eight equivalent $BaO_{12}$ cuboctahedra, faces with two equivalent $BaO_{12}$ cuboctahedra, faces with four equivalent $SrO_{12}$ cuboctahedra and faces with eight equivalent $FeO_6$ octahedra.

However, in Fig. 3.3b shows that there are three unequal Fe sites: The first Fe site, Fe is bonded to six O atoms to form $FeO_6$ octahedra that share corners with six $FeO_6$ octahedra, faces with four equivalent $BaO_{12}$ cuboctahedra, and faces with four equivalent $SrO_{12}$ cuboctahedra. The second Fe site, Fe is bonded to six O atoms to form $FeO_6$ octahedra that share corners with six $FeO_6$ octahedra and faces with eight

**Fig. 3.2** **a** Crystal structures
of $BaFeO_3$. Red spheres is
oxygen, brown sphere is iron
and green cuboctahedra is
$[BaO_{12}]$. **b** Crystal structures
of $BaFeO_3$. Red spheres is
oxygen, green spheres is
barium and brown octahedral
is $[FeO_6]$

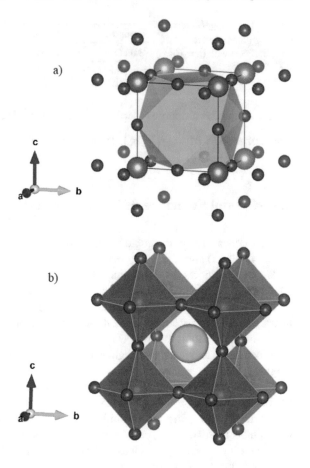

$BaO_{12}$ cuboctahedra. The third Fe site, Fe is bonded to six O atoms to form $FeO_6$
octahedra that share corners with six $FeO_6$ octahedra and faces with eight equivalent
$BaO_{12}$ cuboctahedra. Octahedra sharing corners are not slanted.

## 3.4    Electronic Properties of Ferrite Perovskite

The total density of state (DOS) as function of the energy projected between −
8 and 3.5 eV for $Ba_{1-x}Sr_xFeO_3$, calculated by the GGA approximation as shown in
Fig. 3.4a–b. This figure shows that both compounds have a ground state ferromagnetic
order [25]. The $BaFeO_3$ compound has a half-metallic character. This result is in good
agreement with the theoretical and experimental results [20, 26]. While the compound
$Ba_{0.8}Sr_{0.2}FeO_3$ has a character close to half-metallicity.

Generally there is an underestimation of the GGA approach, hence the need to
seek a more precise approach or introduce a correction to the calculated gap. We

**Fig. 3.3  a** Crystal structures of $Ba_{0.8}Sr_{0.2}FeO_3$. Red spheres is oxygen, brown sphere is iron, green cuboctahedra is [$BaO_{12}$] and blue cuboctahedra is [$SrO_{12}$]. **b** Crystal structures of $Ba_{0.8}Sr_{0.2}FeO_3$. Red spheres is oxygen, green spheres is barium and brown octahedral is [$FeO_6$]

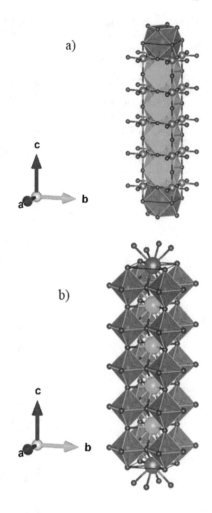

chose the mBJ approach as a first correction and the GGA + U approach as a second correction applied to these compounds.

Figure 3.4c–j illustrates the total and partial DOS calculated using the GGA + U (U = 4.5 and 6 eV) and TB-mBJ approximations. The total DOS profiles are quite similar at the Fermi level for both GGA + U and TB-mBJ approximations and exhibit 100% spin polarization at the Fermi level. This strong polarization due to half of the metallicity. Moreover, Fe-3d and O-2p atoms have a significant contribution to the total density near the Fermi level. This confirms that significant hybridization occurs between the O-2p and Fe-3d orbitals [27]. This suggests that the half-metallicity of BaFeO3 is mainly due to the 2p(O)-3d(Fe) coupling. The Fe d orbital is divided into eg doubly degenerate and triply degenerate t2g states. Figure 3.2 shows that there is strong hybridization between the Fe-eg and O-p states, while the Fe-t2g and O-p states are weakly hybridized. The strong coupling between the Fe-eg and O-p states

**Fig. 3.4  a–b** Total and
partial density of states of
$Ba_{1-x}Sr_xFeO_3$ with (x = 0
and x = 0.2) using GGA
approach. **c–d** Total and
partial density of states of
$Ba_{1-x}Sr_xFeO_3$ with (x = 0
and x = 0.2) using TB-mBJ
approach. **e –j** Total and
partial density of states of
$Ba_{1-x}Sr_xFeO_3$ with (x = 0
and x = 0.2) using GGA +
U approach

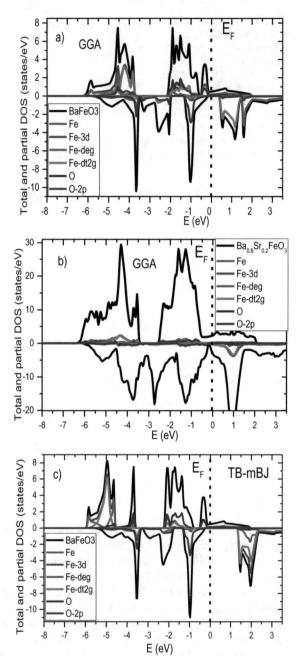

**Fig. 3.4**  (continued)

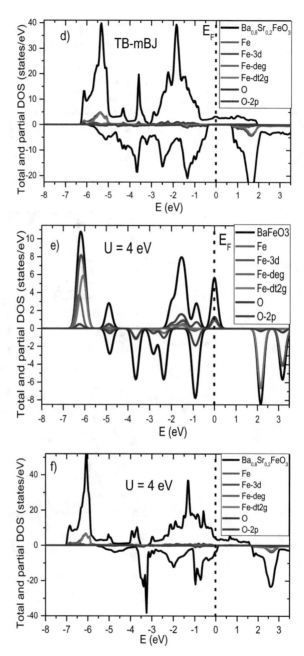

**Fig. 3.4**  (continued)

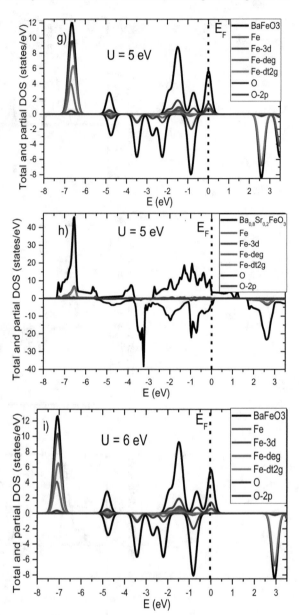

**Fig. 3.4** (continued)

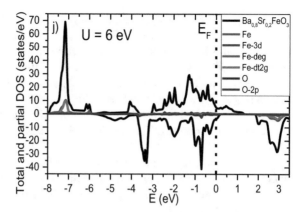

is due to the direct overlap of these orbitals. The $t_{2g}$ states contributed weakly to the spin up states at the Fermi level.

To know the nature of the ground state of the $Ba_{1-x}Sr_xFeO_3$ perovskite with ($x = 0.0.2$), we calculated the energy difference $\Delta E = (E_{AFM} - E_{FM})$ where $E_{AFM}$ and $E_{FM}$ are the energies of the Fe atoms with AFM and FM coupling. Table 3.1 shows that $\Delta E$ are positive, which underlines that the ground state FM is more stable than the AFM declares [28]. The value of the difference energy decreases when Sr is substituted in Ba. This can be explained by the distance. If the distance between atoms is wide, it will be AFM coupling [29].

## 3.5 Magnetic and Magnetocaloric Properties of Ferrite Perovskites

In order to determine the magnetic properties and magnetocaloric effect of $Ba_{1-x}Sr_xFeO_3$ by Monte Carlo simulations, the nearest-, next-nearest- and next-next-nearest-neighbors exchange interactions $J_1$, $J_2$ and $J_3$, respectively, were calculated by using the DFT method. These values of these couplings are cited in Table 3.2. We have shown that the $J_{ij}$ values were decreased during $BaFeO_3$ doping with a concentration of 0.2. This is due to a decrease in the value of the energy difference $\Delta E$ and the magnetic spin moment.

To study the magnetic and electronic properties with application of U and TB-mBJ lead to increase the total magnetic moment, as well as the moment magnetic of Fe [30]. Table 3.3 shows the magnetic spin moment of iron calculated using the mBJ potential in combination with the GGA-PBE approximation and the GGA + U approximation for $Ba_{1-x}Sr_xFeO_3$ ($x = 0, 0.2$). The magnetic spin moment of iron in the compound $BaFeO_3$ calculated by GGA + U with U = 4 eV is 3.89272 $\mu_B$, U = 5 eV is 4.03061 $\mu_B$ and U = 6 eV is 4.11098 $\mu_B$. These values are the same as the values measured theoretically: 4 $\mu_B$ (U = 4 eV), 4.03 $\mu_B$, 4 $\mu_B$ (U = 5 eV) and 3.97 $\mu_B$ (U = 6 eV) [19, 20, 31, 32]. However, the magnetic spin moment of iron calculated using TB-mBJ is 3.67282 $\mu_B$. This value is very close to that of the experimental 3.5 $\mu_B$ measured by N. Hayashi [33]. The magnetic spin moment of iron in $Ba_{0.8}Sr_{0.2}FeO_3$ calculated by GGA + U with U = 4 eV is 3.75290 $\mu_B$, U = 5 eV is 3.87432 $\mu_B$ and U = 6 eV is 3.97222 $\mu_B$. While the magnetic spin moment of iron in $Ba_{0.8}Sr_{0.2}FeO_3$ calculated by TB-mBJ approximation is 3.59335. From these results we conclude that the substitution of Sr in Ba leads to the decrease in the value of the magnetic moment. This result has been shown experimentally by K Yoshii [12].

Figure 3.5a displays the magnetization as a function of the temperature measured under a H = 1, 3 and 5T magnetic field of $BaFeO_3$. This magnetization curve presents a phase transition from a ferromagnetic state (FM) to a paramagnetic state (PM) near the Curie temperature $T_C$.

We obtained a value of $T_c$ equal to 102 k and 53 K and for x = 0, 0.2 respectively. This value of the temperature $T_C$ is determined from the curve dM/dT as shown in Fig. 3.5b. Tc value decrease with Sr substitution. The computed results are in good agreement with available experimental [12].

Figures 3.6 and 3.7, show the susceptibility $\chi$ and the specific heat as a function of the temperature under the influence of the magnetic field 0-5T and the substitution Sr.

**Table 3.3** The total magnetic moment calculated with GGA, GGA + U and with TB-mBJ approach were compared with experimental results and other theoretical results $Ba_{1-x}Sr_xFeO_3$ with (x = 0, 0.2)

| Compounds | Total magnetic moment of Fe atom m($u_B$) | | | | | |
|---|---|---|---|---|---|---|
| | Theoretical GGA | Theoretical GGA + U | | | Theoretical TB-mBJ | Experimental |
| | | U = 4 eV | U = 5 eV | U = 6 eV | | |
| $BaFeO_3$ | 3.24098 | 3.89272 | 4.03061 | 4.11098 | 3.67282 | 3.5 [33] |
| $Ba_{0.8}Sr_{0.2}FeO_3$ | 3.02102 | 3.75290 | 3.87432 | 3.97222 | 3.59335 | – |
| $BaFeO_3$ [20] | 3.18 | – | – | 3.97 | – | – |
| $BaFeO_3$ [19] | 3.65 | – | 4.03 | | – | – |

**Table 3.4** Summary of the magnetic entropy obtained by the Monte Carlo method of $Ba_{1-x}Sr_xFeO_3$ with (x = 0, 0.2) compared to experimental results

| Perovskite | H (T) | $T_c$ (K) | $|\Delta S^{max}|$ (J/kg.K) | References |
|---|---|---|---|---|
| $BaFeO_3$ | 5 | 102 | 5.5 | This Work |
| $Ba_{0.8}Sr_{0.2}FeO_3$ | 5 | 53 | 4 | This Work |
| $BaFeO_3$ | 5 | 111 | 5.8 | [34] |
| $Ba_{0.8}Sr_{0.2}FeO_3$ | 5 | 55 | 4.9 | [12] |
| $GdFeO_3$ | 5 | 2.5 | 35 | [35] |
| $TmFeO_3$ | 5 | 17 | 9.01 | [36] |
| $TbFeO_3$ | 5 | 9 | 17.42 | [36] |

**Fig. 3.5  a** Magnetization as a function of temperatures with different external magnetic fields 1T, 3T and 5 T for $Ba_{1-x}Sr_xFeO_3$. **b** dM/dT as a function of temperatures with different external magnetic fields H = 1, 3 and 5 T for $Ba_{1-x}Sr_xFeO_3$

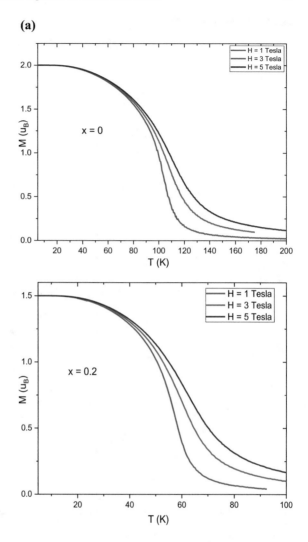

**Fig. 3.5** (continued)

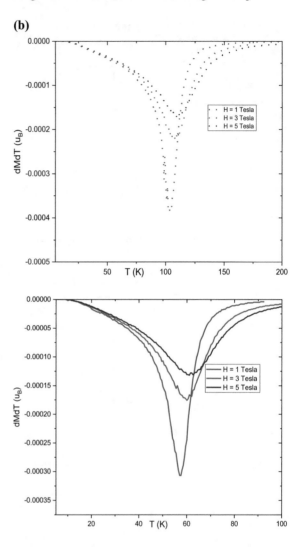

We have noticed that the maximum of the maximum susceptibility $\chi$ and the specific heat are strongly decrease with the increase of the magnetic field while preserving the value of the critical temperature $T_c$.

Regarding the substitution Sr, as shown in the two figures, the maximum of the susceptibility and the specific heat decreased when we substituted Sr in Ba by a concentration of 0.2.

Figure 3.8 shows the reciprocal change in magnetic entropy as a function of temperature in various magnetic fields applied of $Ba_{1-x}Sr_xFeO_3$. As expected, the maximum of the magnetic entropy increases with the applied magnetic field. The curves $-\Delta S_m$ (T) have an increasing shape for temperatures lower than the Curie

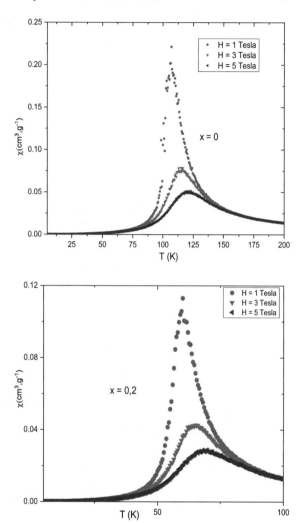

**Fig. 3.6** Susceptibility versus temperatures with different external magnetic fields H = 1T, 3T and 5 T for $Ba_{1-x}Sr_xFeO_3$

temperature, then reach maximum values for $T = T_C$. The maximum value of $-\Delta S_m$ (T) for x = 0 and x = 0.2 are 5.5 J.K$^{-1}$.kg$^{-1}$ and 4 J.K$^{-1}$.kg$^{-1}$ respectively around Tc for a magnetic field change from 0 to 5 T. The Sr substitution led to decrease the maximum value of $-\Delta S_m$ (T).

This value is in good agreement with the experimental results with a small difference in the maximum value of the magnetic entropy as shown in the third table (Table 3.4).

**Fig. 3.7** Specific heat of
BSFO under different
magnetic fields. H = 1, 3 and
5 T for $Ba_{1-x}Sr_xFeO_3$

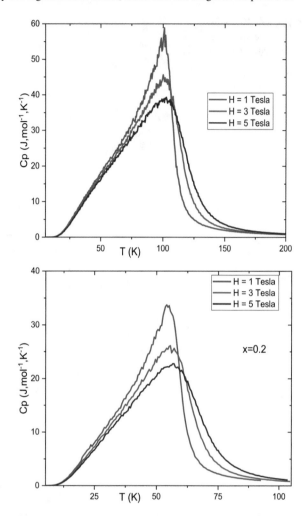

## 3.6   Conclusions

In this chapter, we have tried to review the Monte Carlo and ab initio methods, the
basic theories and then the approaches that we used in our calculation. In a first case
we used the GGA approach implemented on Wien2k to calculate the electronic and
magnetic properties. The preliminary results confronted with the experimental one,
the latter is generally due to the fact that the term of exchange correlation is not known
exactly. It was necessary to use other corrective approaches, namely GGA + U, TB-
mBJ, to have great precision on the results obtained, but the use of one approach
or the other depends on the system, nature of the atoms, electronic configuration.
Correction with TB-mBJ was generally successful on magnetic and non-magnetic
materials. But in magnetic materials we cannot locate the f or d bands with this

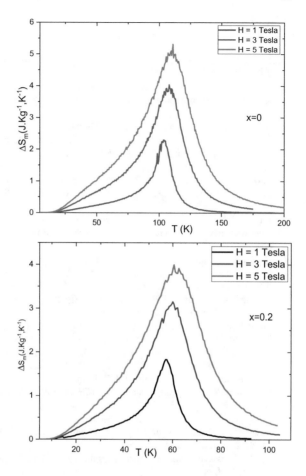

**Fig. 3.8** Temperature dependence of the magnetic entropy change under a magnetic-field for $Ba_{1-x}Sr_xFeO_3$

approach. On the other hand, the GGA + U approximation the exact choice for the value of U allows a localization of the bands (d or f), a total moment, and a fixed load. With the two approaches (mBJ and GGA + U) we manage to correct the problems and we have obtained results in good agreement with those found experimentally.

The main objective of this chapter was the theoretical study of the Sr effect on electronic, magnetic and magnetocaloric properties in $BaFeO_3$ compound. Analysis of total and partial DOS and the value of difference energy showed that the system $Ba_{1-x}Sr_xFeO_3$ has a ferromagnetic behavior and half metallic character. The Sr substituted $BaFeO_3$ leads to decrease the magnetic moment and Curie temperature. Moreover, we used the Monte Carlo simulation to study the magnetic properties such as magnetization, magnetic entropy, susceptibility and specific heat. The maximum of these magnetic properties is decrease with Sr substitution. The $Ba_{0.8}Sr_{0.2}FeO_3$ material showed the second-order PM-FM phase transition around 53 K with a large magnetic entropy $4\,J.K^{-1}.kg^{-1}$ for x = 0.2 at 5T. Our results could show that BFSO has desirable magnetic properties that meet the needs of magnetic/spintronic devices.

# References

1. J.N. Kuhn, U.S. Ozkan, Surface properties of Sr-and Co-doped LaFeO$_3$. J. Catal.Catal. **253**(1), 200–211 (2008)
2. K. Mori, T. Kamiyama, H. Kobayashi, K. Oikawa, T. Otomo, S. Keda, Structural evidence for the charge disproportionation of Fe$^{4+}$ in BaFeO$_{3-\delta}$. J. Phys. Soc. Japan **72**(8), 2024–2028 (2003)
3. A. Ruffo, M.C. Mozzati, B. Albini, P. Galinetto, M. Bini, Role of non-magnetic dopants (Ca, Mg) in GdFeO$_3$ perovskite nanoparticles obtained by different synthetic methods: structural, morphological and magnetic properties. J. Mater. Sci. Mater. Electron. **31**(20), 18263–18277 (2020)
4. S.M. Yakout, H.A. Mousa, H.T. Handal, W. Sharmoukh, Role of non-magnetic dopants on the room temperature ferromagnetism and optical properties of BaSnO$_3$ perovskite. J. Solid-State Chem. **281**, 121028 (2020)
5. I. Elahi, W. Akbar, M. Zulfiqar, S. Nazir, Thermodynamic stability and creation of large half-metallic gap in BaZrO$_3$ via non-magnetic elements doping. J. Phys. Chem. Solids **164**, 110616 (2022)
6. C. Wang, Y. Liu, Y. Lu, P. Wu, W. Zhou, The electronic structure and room temperature ferromagnetism in non-magnetic element X (X= Al, Mg and Li) doped SrSnO$_3$ from hybrid functional calculations. Comput. Mater. Sci.. Mater. Sci. **145**, 102–108 (2018)
7. A. Sagdeo, K. Gautam, P.R. Sagdeo, M.N. Singh, S.M. Gupta, A.K. Nigam, A. Chakrabarti, Large dielectric permittivity and possible correlation between magnetic and dielectric properties in bulk BaFeO$_{3-\delta}$. Appl. Phys. Lett. **105**(4), 042906 (2014)
8. T. Matsui, H. Tanaka, N. Fujimura, T. Ito, H. Mabuchi, K. Morii, Structural, dielectric, and magnetic properties of epitaxially grown BaFeO$_3$ thin films on (100) SrTiO$_3$ single-crystal substrates. Appl. Phys. Lett. **81**(15), 2764–2766 (2002)
9. A.S. Kumar, P.D. Babu, P. Srinivas, A.K. Bhatnagar, Weakening of the helical antiferromagnetic order in BaFeO$_{3-\delta}$ nanoparticles due to Sr-Doping. J. Supercond. Novel Magn.Supercond. Novel Magn. **30**(9), 2563–2568 (2017)
10. C. Berger, E. Bucher, A. Windischbacher, A.D. Boese, W. Sitte, Strontium-free rare earth perovskite ferrites with fast oxygen exchange kinetics: experiment and theory. J. Solid-State Chem. **259**, 57–66 (2018)
11. B. Bhushan, A. Basumallick, N.Y. Vasanthacharya, S. Kumar, D. Das, Sr induced modification of structural, optical and magnetic properties in Bi$_{1-x}$Sr$_x$FeO$_3$ (x= 0, 0.01, 0.03, 0.05 and 0.07) multiferroic nanoparticles. Solid State Sci. **12**(7), 1063–1069 (2010)
12. K. Yoshii, N. Hayashi, M. Mizumaki, M. Takano, Magnetocaloric effect of Sr-substituted BaFeO$_3$ in the liquid nitrogen and natural gas temperature regions. AIP Adv. **7**(4), 045117 (2017)
13. R. Terki, H. Feraoun, G. Bertrand, H. Aourag, Full potential linearized augmented plane wave investigations of structural and electronic properties of pyrochlore systems. J. Appl. Phys. **96**(11), 6482–6487 (2004)
14. B. Traoré, G. Bouder, W. Lafargue-Dit-Hauret, X. Rocquefelte, C. Katan, F. Tran, M. Kepenekian, Efficient and accurate calculation of band gaps of halide perovskites with the Tran-Blaha modified Becke-Johnson potential. Phys. Rev. B **99**(3), 035139 (2019)
15. B. Himmetoglu, A. Floris, S. De Gironcoli, M. Cococcioni, Hubbard-corrected DFT energy functionals: the LDA+ U description of correlated systems. Int. J. Quantum Chem. **114**(1), 14–49 (2014)
16. D. Koller, F. Tran, P. Blaha, Improving the modified Becke-Johnson exchange potential. Phys. Rev. B **85**(15), 155109 (2012)
17. A.D. Becke, M.R. Roussel, Phys. Rev. A **39**, 3761 (1989)
18. G. Rahman, S. Sarwar, Phys. Status Solidi B **253**, 329–334 (2016)
19. M. Mizumaki, H. Fujii, K. Yoshii, N. Hayashi, T. Saito, Y. Shimakawa, T. Uozumi, M. Takano, Phys. Status Solidi C **12**(6), 818–821 (2015)

20. H. Noura, GGA+ U-DFT+ U modeling structural, electronic and magnetic properties investigation on the ferromagnetic and anti-ferromagnetic $BaFeO_3$ characteristics: insights from First-principle calculation. Superlattices Microstruct. Microstruct. **76**, 425–435 (2014)

21. M. Taib, N.H. Hussin, M.H. Samat, O.H. Hassan, M.Z.A. Yahya, Structural, electronic and optical properties of $BaTiO_3$ and $BaFeO_3$ from first principles LDA+ U study. Int. J. Electroact Mater **4**, 14–17 (2016)

22. G. Kadim, R. Masrour, A. Jabar, E.K. Hlil, Room-temperature large magnetocaloric, electronic and magnetic properties in $La_{0.75}Sr_{0.25}MnO_3$ manganite: Ab initio calculations and Monte Carlo simulations. Phys. A Stat. Mech. Appl. **573**, 125936 (2021)

23. A.M. Ferrenberg, R.H. Swendsen, New Monte Carlo technique for studying phase transitions. Phys. Rev. lett. **61**(23), 2635 (1988)

24. G. Kadim, R. Masrour, A. Jabar, Large magnetocaloric effect, magnetic and electronic properties in $Ho_3Pd_2$ compound: Ab initio calculations and Monte Carlo simulations. J. Magn. Magn. Mater.Magn. Magn. Mater. **499**, 166263 (2020)

25. H.J. Feng, F.M. Liu, Electronic structure of $BaFeO_3$: an abinitio DFT study (2007). arXiv: 0704.2985

26. Z. Li, T. Iitaka, T. Tohyama, Pressure-induced ferromagnetism in cubic perovskite $SrFeO_3$ and $BaFeO_3$. Phys. Rev. B **86**(9), 094422 (2012)

27. R.A. Jishi, H.M. Alyahyaei, Effect of hybridization on structural and magnetic properties of iron-based superconductors. New J. Phys. **11**(8), 083030

28. Z. Wu, J. Yu, S. Yuan. Strain-tunable magnetic and electronic properties of monolayer $CrI_3$. Phys. Chem. Chem. Phys. (2019)

29. A. Ramasubramaniam, D. Naveh, Mn-doped monolayer $MoS_2$: An atomically thin dilute magnetic semiconductor. Phys. Rev. B **87**, 195201 (2013)

30. B. Fadila, M. Ameri, D. Bensaid, M. Noureddine, I. Ameri, S. Mesbah, Y. Al-Douri, Structural, magnetic, electronic and mechanical properties of full-Heusler alloys $Co_2YAl$ (Y= Fe, Ti): first principles calculations with different exchange-correlation potentials. J. Magn. Magn. Mater.Magn. Magn. Mater. **448**, 208–220 (2018)

31. I. Cherair, N. Iles, L. Rabahi, A. Kellou, Effects of Fe substitution by Nb on physical properties of $BaFeO_3$: A DFT+ U study. Comput. Mater. Sci.. Mater. Sci. **126**, 491–502 (2017)

32. I. Cherair, E. Bousquet, M.M. Schmitt, N. Iles & A. Kellou. Strain induced Jahn-Teller distortions in $BaFeO_3$: A first-principles study (2017). arXiv:1710.10997

33. N. Hayashi, T. Yamamoto, H. Kageyama, M. Nishi, Y. Watanabe, T. Kawakami, Y. Matsushita, A. Fujimori, M. Takano, Angew. Chem. Int. Ed.. Chem. Int. Ed. **50**, 12547 (2011)

34. M. Mizumaki, K. Yoshii, N. Hayashi, T. Saito, Y. Shimakawa, M. Takano, Magnetocaloric effect of field-induced ferromagnet $BaFeO_3$. J. Appl. Phys. **114**(7), 073901 (2013)

35. M. Das, S. Roy, P. Mandal, Giant reversible magnetocaloric effect in a multiferroic $GdFeO_3$ single crystal. Phys. Rev. B **96**(17), 174405 (2017)

36. Y.J. Ke, X.Q. Zhang, Y. Ma, Z.H. Cheng, Anisotropic magnetic entropy change in $RFeO_3$ single crystals (R= Tb, Tm, or Y). Sci. Rep. **6**, 19775 (2016)

# Chapter 4
# Magnetic and Magnetocaloric, Electronic, Magneto-Optical, and Thermoelectric Properties of Perovskite Chromites

**Abstract** In this study, we employed a combination of density functional theory (DFT) and Monte Carlo simulations to comprehensively investigate the multifaceted properties of the GdCrO$_3$ perovskite. Our analysis encompassed magnetocaloric effects, electronic structures, optical properties, thermoelectric characteristics, and magnetic behavior.

## 4.1 Introduction

Chromite of ACrO$_3$ perovskite structure with A is a rare earth, La, Ho, Er, Pr, Gd..., crystallize in an orthorhombic structure [1–4]. Rare-earth chromites are typify by the coexistence of magnetoelectric multiferroics whose ferroelectricity and ferromagnetism simultaneous development under conditions of ambient temperature and pressure [5]. On the other hand, these properties make this material a candidate for many technological applications, in particular optoelectronic, spintronic, and thermoelectric applications [6, 7], which is envisaged in the field of information storage (magnetic memories and heads readings) or the development of electrical components (temperature reversible circuit breakers) and the manufacture of spintronic devices [8–13].

Chromium-based materials are generally antiferromagnetic (AFM) below a critical temperature called Néel temperature and paramagnetic (PM) above Néel temperature.

In this chapter, we present the various results of our calculations. Therefore, we applied two methods. The first method is the density functional theory which is related to the study of the electronic structure [14]. The second method is the Monte Carlo simulation, which is related to the study of the magnetocaloric effect [15].

We mention that our calculation was performed by dealing with electronic, optical, thermoelectric, magnetic and magnetic properties of the GdCrO$_3$ system.

© The Author(s), under exclusive license to Springer Nature Switzerland AG 2024
R. Masrour, *Magnetoelectronic, Optical, and Thermoelectric Properties of Perovskite Materials*, SpringerBriefs in Materials,
https://doi.org/10.1007/978-3-031-48967-9_4

## 4.2   Calculation Methods

### 4.2.1   Density Functional Theory

Our calculations were processed using the density functional theory DFT approach applied in WIEN2k. The exchange–correlation potential used to calculate the electronic and optical properties is the generalized gradient approximation (GGA) parametrized by perdew, Burke and Ernzerhof [16].

The first step of this work consists in specifying the values of the important parameters, which influence the time and the precision of the calculation. For this, tests of convergence of the total energy must be carried out according to two parameters:

- Basic functions are performed up to $l_{max} = 10$.
- The atomic radii of the muffin tin $R_{MT}$ spheres we used are: $R_{MT}(Gd) = 2.28$, $R_{MT}(Cr) = 1.97$ and $R_{MT}(O) = 1.78$.
- The number of k points in the first Brillouin zone were based on the 11 * 10 * 7 Monkhorst–Pack scheme.
- A choice on the cutoff energy $E_{cut} = -6\,eV$ to ensure the convergence of the total energy $E_T$
- Convergence of the charge has been obtained when the difference is less than 1 mRy.
- The parameter of $G_{max} = 12$ is the norm of the largest wave vector used for the plane wave development of the charge density.

On the other hand, for the optical properties, we used 1000 k points in the irreducible BZ to obtain a more precise convergence.

Furthermore, thermoelectric properties are obtained from the electronic structure calculated using the program BoltzTraP, we used 1000 k-points in the irreducible BZ in order to obtain a convergence more precise.

After initializing the data, we get the total energy through a self-consistent calculation. This allows us to subsequently determine the physical properties of $GdCrO_3$.

DFT calculations are performed at zero temperature. However, it is possible to obtain an estimate of the Neel temperature for antiferromagnetic materials using the Ising model.

### 4.2.2   Monte Carlo Study

The Hamiltonian of the mixed-spin Ising model of $GdCrO_3$ is:

$$H = -J_{CrCr} \sum_{<i,j>} S_i S_j - J_{GdGd} \sum_{<k,l>} \sigma_k \sigma_l - J_{CrGd}$$

$$\sum_{<i,k>} S_i \sigma_k - h \left( \sum_i S_i + \sum_k \sigma_k \right)$$

This Hamiltonian consists of four sums. The first, second and third concern the interactions between the adjacent spins denoted <i, j>, <k, l> and <i, k> which means that the sites i, j, k, and l are the nearest neighbors. The fourth sum refers to the energy resulting from an external magnetic field.

Conventionally, the Hamiltonian function contains only signs (−) on its terms. This convention allowed the classification of the Ising model depending on the sign of the exchange coupling. Therefore, for:

- $J_{ij} > 0$, the interaction between the $S_i$, $S_j$, $\sigma_k$ and $\sigma_l$, spins are of the ferromagnetic type.
- $J_{ij} < 0$, the interaction between the $S_i$, $S_j$, $\sigma_k$ and $\sigma_l$, spins are of the antiferromagnetic type.
- $J_{ij} = 0$, there is no interaction between the spins $S_i$, $S_j$, $\sigma_k$ and $\sigma_l$.

Therefore, the Ising model is ferromagnetic if the configuration for which the nearest neighboring spins have the same polarization direction has a higher probability. Contrary, the model is antiferromagnetic if the majority of the nearest neighbor spins are in an antiparallel configuration.

Otherwise, the convention in the second term of the Hamiltonian provides how a spin interacts with the external magnetic field. Therefore, for:

- $H > 0$, the Si spin has the direction of the external magnetic field.
- $H < 0$, the Si spin has the opposite direction of the external magnetic field.
- $H = 0$, the Si spin is not influenced by the external magnetic field.
- $S_{Cr} = 3/2$ and $\sigma_{Gd} = 7/2$ the spin of Chromite and Gadolinium respectively.
- $J_{CrCr}$ is the exchange constant between the first couplings of the spins Cr at the sites $i$ and $j$.
- $J_{GdGd}$ is the exchange constant between the first couplings of the spins Gd at the sites k and l.
- The number of Gd and Cr atoms in the structure (Fig. 4.1) are $N_{Gd} = 2210$, $N_{Cr} = 2420$ respectively.

To find the value of the critical temperature using Monte Carlo simulation it's necessary to calculate the coupling constants $J_{GdGd}$, $J_{CrGd}$ and $J_{CrCr}$. We have calculated them from the difference energy ΔE and we have quoted the values of this coupling constants in Table 4.2.

These magnetic properties are defined in the following [17]:

The internal energy per site E is:

$$E = \frac{1}{N} \langle H \rangle$$

The thermal magnetizations are:

$$M_S = \frac{1}{N_s} \langle \sum_i S_i \rangle$$

$$M_\sigma = \frac{1}{N_\sigma} \langle \sum_i \sigma_i \rangle$$

The total thermal magnetization of GdCrO$_3$ is:

$$M_{Total} = \frac{N_s M_S + N_\sigma M_\sigma}{N_s + N_\sigma}$$

The magnetic susceptibilities are given by:

$$\chi_S = \beta \left( \langle M_S^2 \rangle - \langle M_S \rangle^2 \right)$$

$$\chi_\sigma = \beta \left( \langle M_\sigma^2 \rangle - \langle M_\sigma \rangle^2 \right)$$

where $\beta = \frac{1}{k_B T}$ with and k$_B$ is the Boltzmann's constant.
The total magnetic susceptibility is:

$$\chi_{Total} = \frac{N_s \chi_S + N_\sigma \chi_\sigma}{N_s + N_\sigma}$$

The magnetic field-induced entropy change is given by:

$$\Delta S(T, h) = \int_0^h \left( \frac{\partial M}{\partial T} \right)_h dh$$

The RCP is given by:

$$RCP = \int_{T_C}^{T_h} \Delta S(T) dT$$

where T$_c$ and T$_h$ are the cold and the hot temperatures corresponding to both ends of the half-maximum value of $\Delta$S(T) respectively.

## 4.3 Crystal Structure of Perovskite Chromites

The compound $GdCrO_3$ crystallizes under ambient conditions in the orthorhombic structure (Fig. 4.1) with a Pnma space group. The conventional unit cell of the orthorhombic structure contains four Chromite atoms, four Gadolinium atoms and 12 Oxygen atoms occupying the following positions Table 4.1.

The lattice parameters of $GdCrO_3$ are a = 5.301 (Å), b = 5.533 (Å), c = 7.599 (Å) and $\alpha = \beta = \gamma = 90°$ [18].

$Gd^{3+}$ is bonded in an eight-coordinate geometry to eight $O^{2-}$ atoms. $Cr^{3+}$ is bonded to six $O^{2-}$ atoms to form corner-sharing $CrO_6$ octahedra (Fig. 4.2).

There are two inequivalent $O^{2-}$ sites. In the first $O^{2-}$ site, $O^{2-}$ is bonded in a five-coordinate geometry to three equivalent $Gd^{3+}$ and two equivalent $Cr^{3+}$ atoms. In the second $O^{2-}$ site, $O^{2-}$ is bonded to two equivalent $Gd^{3+}$ and two equivalent $Cr^{3+}$ atoms to form distorted corner-sharing $OGd_2Cr_2$ tetrahedra.

**Fig. 4.1** Structure of $GdCrO_3$ compound

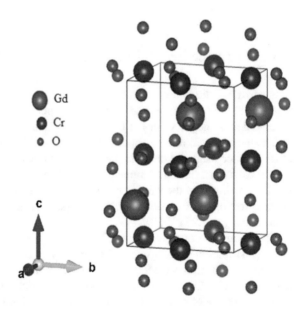

**Table 4.1** Atomic positions of $GdCrO_3$

| Wyckoff | Element | x | y | z |
|---------|---------|------|------|------|
| 4b | Cr | 0 | 1/2 | 1/2 |
| 4c | Gd | 0.02 | 0.94 | 3/4 |
| 4c | O | 0.1 | 0.46 | 1/4 |
| 8d | O | 0.7 | 0.3 | 0.45 |

**Fig. 4.2** Crystal structures of GdCrO₃. Red spheres is oxygen, pink sphere is Gadolinium and blue octahedra is [CrO₆]

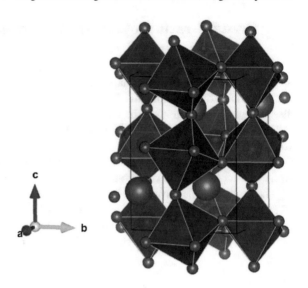

## 4.4   Electronic Properties of Perovskite Chromites

In this part, we are interested in the calculation of band structures. The calculations were carried out along the high symmetry directions in the first Brillouin zone in order to determine the values of the energy gaps of GdCrO₃.

The fundamental energy gap is defined as the difference between the maximum of the valence band and the minimum of the conduction band.

The calculated band structures for GdCrO₃ are presented in Fig. 4.3. We notice that the maximum of the valence band and the minimum of the conduction band are at the same point $\Gamma$. Consequently, this compound is direct gap semiconductors $\Gamma$–$\Gamma$ with a band gap 1.26 eV [19].

The electronic density of states (DOS) is an essential quantity for calculating the energy distribution of electrons in the valence and conduction bands.

Figure 4.4 illustrates the total and partial densities of states of GdCrO₃ compound obtained by the GGA-PBE approximation. The partial densities of states are deduced from the density of states projected onto atomic orbitals of this compound. There is symmetry in TDOS with respect to the energy axis. This confirms that the total magnetic moment equals 0 $\mu_B$. This can be attributed that the structure of GdCrO₃ is antiferromagnetic (G-AFM) type.

The PDOS discussion reveals that this is a strong electronic contribution of the O-p and Cr-d orbitals around fermi and Gd-f and O-p in [−4, −2].

This result was shown by calculating the energy difference $\Delta E = (E_{AFM} - E_{FM})$, where $E_{AFM}$ and $E_{FM}$ represent the energies of the atoms of Gd and Cr with FM and AFM coupling (Fig. 4.5).

We obtain the negative values of $\Delta E$ (Fig. 4.5d), which confirms that the ground state AFM is more stable than FM.

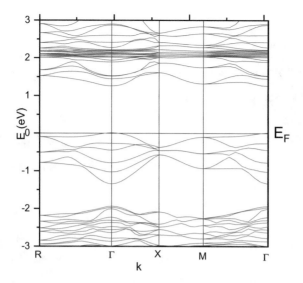

**Fig. 4.3** Energy band structure of GdCrO$_3$

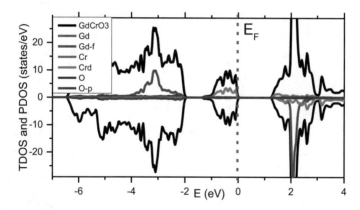

**Fig. 4.4** Total and Partial DOS calculated of GdCrO$_3$ compound

## 4.5 Dielectric and Optical Properties of Perovskite Chromites

In this part, we will present the optical properties of GdCrO$_3$. Namely the dielectric function, absorption and reflectivity using the ab initio methods implemented in the Wien2k code using the GGA-PBE approximation.

The optical properties of GdCrO$_3$ are expressed by the dielectric frequency function which is written in the following form:

**Fig. 4.5** Three antiferromagnetic ordering configurations (**a–c**) in orthorhombic GdCrO$_3$: **a** C-AFM, **b** A-AFM and **c** G-AFM. The arrows indicate magnetic moment orientations on Gd and Cr atoms. **d** The difference energy ($\Delta E$) between E$_{AFM}$ and E$_{FM}$

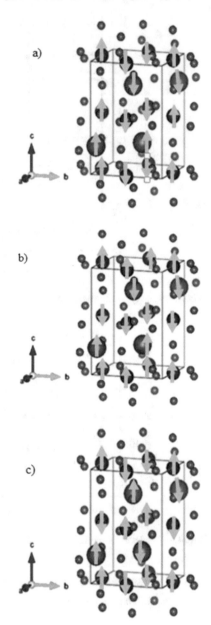

**Fig. 4.5**  (continued)

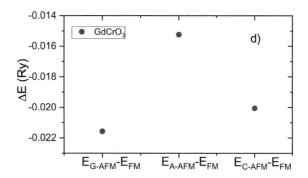

$$\varepsilon(\omega) = \varepsilon_1(\omega) + i\varepsilon_2(\omega)$$

where $\varepsilon_2(\omega)$ the imaginary part of the dielectric function it was calculated from relation [20]:

$$\varepsilon_2(\omega) = \frac{4\pi e^2}{\omega^2 m^2} \sum_{i,j} \int <i|M|j>^2 f i (1 - f j) * \delta(Ef - Ei - \omega) d^3 k$$

where e is the electron charge, m it's mass and M is the dipole matrix element, i and j are the initial and final states, respectively, fi is the Fermi distribution function. $E_i$ is the energy of electron.

$\varepsilon_1(\omega)$ the real part of the dielectric function can be extracted from the imaginary part of the dielectric function $\varepsilon_2(\omega)$:

$$\varepsilon_1(1) = Re(\varepsilon(1)) = 1 + \frac{\pi}{2} P \int_0^\infty \frac{\omega' \varepsilon_2(\omega')}{\omega'^2 - \omega^2} d\omega'$$

The real $\varepsilon_1(\omega)$ and imaginary $\varepsilon_2(\omega)$ parts of the dielectric function for GdCrO$_3$ Compound are shown in Fig. 4.6a, b, as we can see, the real part of the dielectric function is negative in some regions, which means that there is a total reflection in this area. The static dielectric (real part) constant $\varepsilon(0)$ can be extracted from the low energy limit of $\varepsilon(\omega)$. The static dielectric coefficient $\varepsilon_1(0)$ start at 6.21.

In addition, the peaks observed on spectrum real and imaginary of dielectric function are due to transition from valance band to conduction band. Furthermore, these peaks in our compounds are mainly due to O-2p, Gd-f and Cr-3d states electrons of the valence transitions to the Gd-f and Cr-d states of the conduction bands.

The optical absorption $\alpha(\omega)$ are the most important parameter in optical calculations. Moreover, $\alpha(\omega)$ is the most direct method to understand the bandgap and band structure of crystalline and non-crystalline materials [21].

The absorption $\alpha(\omega)$ obtained directly from the following relation [20]

**Fig. 4.6  a** Variation of real $\varepsilon_1(\omega)$ parts of dielectric function for GdCrO$_3$ Compound; **b** Variation of imaginary $\varepsilon_2(\omega)$ parts of dielectric function for GdCrO$_3$ Compound

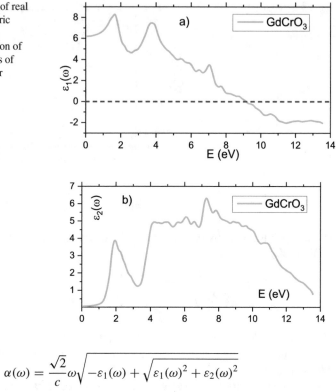

$$\alpha(\omega) = \frac{\sqrt{2}}{c}\omega\sqrt{-\varepsilon_1(\omega) + \sqrt{\varepsilon_1(\omega)^2 + \varepsilon_2(\omega)^2}}$$

However, the absorption of GdCrO$_3$ is presented in Fig. 4.7a. This figure show that the fundamental absorption edge as well as the following the peaks correspond to the direct $\Gamma$–$\Gamma$ transition [22, 23]. The direct optical transition mainly goes from the occupied state of the valence band (VB) Cr-d, O-p to the Gd-f and Cr-d states from 1.30 eV to 4eV.

The spectrum of reflectivity R($\omega$) for a normal incidence on the surface of a crystal, is deduced from the relation [24]

$$R(w) = \left| \frac{\sqrt{\varepsilon(w)} - 1}{\sqrt{\varepsilon(w)} - 1} \right|^2$$

Figure 4.7b display the reflectivity R($\omega$). We can observed that at low energies these compounds have a high reflectivity. It's starts at 0.181%.

**Fig. 4.7 a** Variation of absorption for GdCrO$_3$ Compound **b** Variation of reflectivity for GdCrO$_3$ Compound

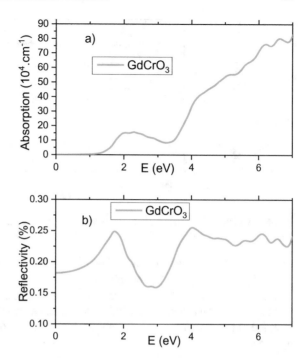

## 4.6 Thermoelectric Properties of Perovskite Chromites

The study of the Seebeck coefficient allows us to determine the efficiency of a thermocouple. The Seebeck coefficient of our compound is shown in Fig. 4.8. This figure show that the S is decrease as the temperature increased. The maximum value of S is $8.75 \times 10^{-6}$ V/K for spin up and spin down.

Figure 4.9 displays the electrical conductivity as function of temperature for spin down. The conductivity response increased with the increase of temperature and near

**Fig. 4.8** Variation of Seebeck coefficient (S) with temperature for GdCrO$_3$

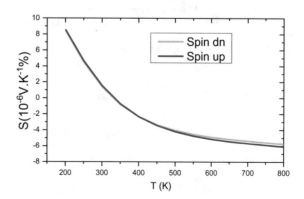

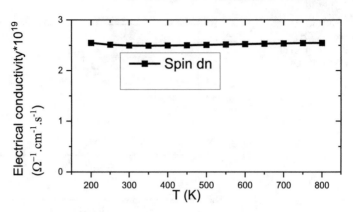

**Fig. 4.9**  Variation of electrical conductivity with temperature for GdCrO₃

about 800 K the conductivity of GdCrO₃ 2.55 * 10¹⁹ ($\Omega^{-1}$ cm$^{-1}$ s$^{-1}$) for spin *dn* and 2.54 * 10¹⁹ ($\Omega^{-1}$ cm$^{-1}$ s$^{-1}$) for spin up.

Figure 4.10 display thermal conductivity as function of temperature. It is observed from the figure that thermal conductivity increased for spin up and spin *dn* with the increase of temperature. The maximum value is 4.90 * 10¹⁴ W K$^{-1}$ m$^{-1}$ s$^{-1}$ for spin up and spin dn.

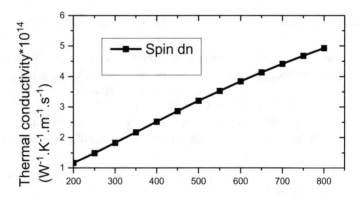

**Fig. 4.10**  Variation of thermal conductivity with temperature for GdCrO₃ Compound

## 4.7 Magnetic and Magnetocaloric Effect of Perovskite Chromites

Magnetic materials have a considerable technological importance, thanks to their great richness of behavior. In addition, magnetism occupies a remarkable place in the description of the fundamental properties of matter. In this part we will deal with the magnetic and magnetocaloric properties of $GdCrO_3$.

The total magnetic moment of Cr and Gd are cited in Table 4.1. They are used as input for Monte Carlo simulations to calculate magnetic properties and magnetocaloric effect (Table 4.2).

To calculate the $T_N$ value using Monte Carlo simulation, the thermal magnetization and susceptibility at h = 0 T obtained by Monte Carlo simulations are presented in Fig. 4.11. From this curve, we deduce that the Neel temperature is equal to $T_N$ = 176 K. This value is equal to that found experimentally [25].

The magnetization curve shows a classical phase transition of the 2nd order at $T_N$ (antiferromagnetic/paramagnetic). We can see that prominent spin reorientation transitions exist in this material. The value of spin reorientation temperature is obtained, $T_{SR}$ = 78 K. The peak in a lower region of spin reorientation transition may be responsible for the long-range spin reorientation of $Fe^{3+}$ sublattice.

Figure 4.12, shows the variation of thermal magnetic entropy change of $GdCrO_3$ obtained by Monte Carlo simulations. The value of $-\Delta S$ increases with increasing

**Table 4.2** The exchange coupling $J_{ij}$ and magnetic moment of magnetic atom for the $GdCrO_3$ compound

| Perovskite | $J_{GdGd}(K)$ | $J_{GdCr}(K)$ | $J_{CrCr}(K)$ | Cr ($\mu_B$) | Gd ($\mu_B$) |
|---|---|---|---|---|---|
| $GdCrO_3$ | +26.020221 | +24.6380937 | −0.4856083 | 2.47035 | 6.9041 |

**Fig. 4.11** The thermal magnetization and susceptibility at h = 0 T obtained using Monte Carlo simulations

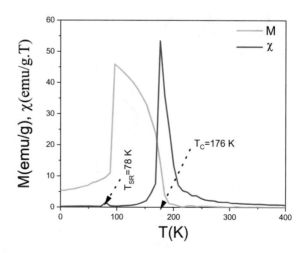

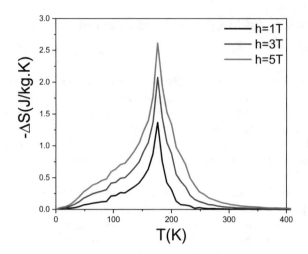

**Fig. 4.12** The thermal magnetic entropy change $-\Delta S$ obtained using Monte Carlo simulations

the values of external magnetic field increases while the critical temperature remains stable. The maximum of magnetic entropy is $-\Delta S = 2.64\,\text{J}\cdot\text{K}^{-1}\cdot\text{kg}^{-1}$ at h = 5 T.

Figure 4.13 displays the variation of relative cooling power with the magnetic field for $GdCrO_3$ using Monte Carlo simulations. RCP varies linearly with magnetic field h. The maximum value of RCP is 25 J/kg is given for h = 5 T.

Figure 4.14 shows, the thermal dependence of relative cooling power for a several magnetic fields. The values of RCP increase with increasing the values of magnetic field for a fixed value of temperature. RCP values increase with increasing magnetic field values and temperature until reached their saturation for each value of magnetic field.

**Fig. 4.13** The field dependence of relative cooling power

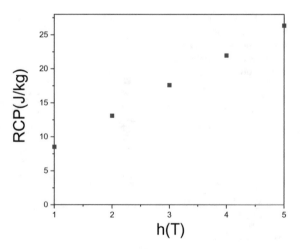

**Fig. 4.14** The thermal dependence of relative cooling power for a several magnetic fields

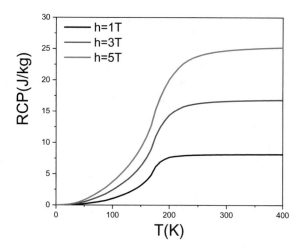

**Fig. 4.15** Magnetic hysteresis cycle for the temperature T = 120, 150 and 190 K

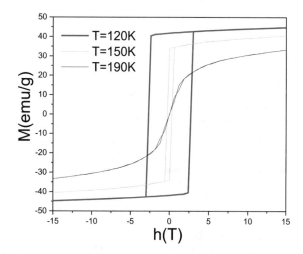

The magnetic hysteresis cycle for several temperatures is given in Fig. 4.15. The magnetic coercive field, and remanent magnetization, decrease with increasing the temperature and become and they became zero above the transition temperature. The system exhibits the superparmagnetic at superior to the transition temperature.

## 4.8 Conclusions

The objective of our research is to look for new materials combining several properties at the same time. Our contribution to the investigation of this type of materials is devoted to perovskite oxides. The investigation of the different properties of these materials was made using an ab-initio method called linearized augmented plane wave method (FP-LAPW) implemented by the Wien2k code and within the framework of the density functional theory (DFT), using GGA approximations and Monte Carlo simulations. In the first part, we studied the electronic, magnetic, thermoelectric properties and finally the magnetic and magnetocaloric properties of perovskite $GdCrO_3$. $GdCrO_3$ crystallize in the Pnma structure with a G-type antiferomagnetic phase. The study of electronic properties: the band structure shows that the compound $GdCrO_3$ is a semiconductor with a direct gap. For the optical properties we have made qualitative studies for certain optical constants, i.e. the dielectric function, the absorption coefficient and the spectrum of reflectivity. The results obtained are predictive and serve as good references for future experimental work. The thermoelectric response is evaluated by calculating electrical conductivity, thermal conductivity and Seebeck coefficient (S) by using BoltzTraP code. The Néel and spin reorientation temperatures of $GdCrO_3$ compound were found using Monte Carlo simulation. The maximum of magnetic entropy increases with increasing the external magnetic field. The values of $h_C$ and $M_r$ decrease with increasing the temperature.

# References

1. A.T. Apostolov, I.N. Apostolova, J.M. Wesselinowa, Microscopic approach to the magneto-electric coupling in $RCrO_3$. Mod. Phys. Lett. B **29**, 1550251 (2015)
2. A.T. Apostolov, I.N. Apostolova, J.M. Wesselinowa, Theory of magnetic field control on polarization in multiferroic $RCrO_3$ compounds. Eur. Phys. J. B. **88**(12), 1–9 (2015)
3. L.H. Yin, J. Yang, P. Tong, X. Luo, C.B. Park, K.W. Shin, Y.P. Sun, Role of rare earth ions in the magnetic, magnetocaloric and magnetoelectric properties of $RCrO_3$ (R= Dy, Nd, Tb, Er) crystals. J. Mater. Chem. C. **4**(47), 11198–11204 (2016)
4. A.K. Zvezdin, Z.V. Gareeva, X.M. Chen, Multiferroic order parameters in rhombic antiferro-magnets $RCrO_3$. J. Phys. Condens. Matter **33**(38), 385801 (2021)
5. N.R. Camara, V. Ta Phuoc, I. Monot-Laffez, M. Zaghrioui, Polarized Raman scattering on single crystals of rare earth orthochromite $RCrO_3$ (R = La, Pr, Nd, and Sm). J. Raman Spectrosc. **48**(12), 1839–1851 (2017)
6. S. Hussain, A.J. Khan, M. Arshad, M.S. Javed, A. Ahmad, S.S.A. Shah, G. Qiao, Charge storage in binder-free 2D-hexagonal CoMoO4 nanosheets as a redox active material for pseudocapacitors. Ceram. Int. **47**(6), 8659–8667 (2021)
7. S. Hussain, M. Hassan, M.S. Javed, A. Shaheen, S.S.A. Shah, M.T. Nazir, G. Liu, Distinctive flower-like $CoNi_2S_4$ nanoneedle arrays (CNS–NAs) for superior supercapacitor electrode performances. Ceram. Int. **46**(16), 25942–25948 (2020)
8. S. Hussain, N. Ullah, Y. Zhang, A. Shaheen, M.S. Javed, L. Lin, G. Qiao, One-step synthesis of unique catalyst Ni9S8@ C for excellent MOR performances. Int. J. Hydrogen Energy **44**(45), 24525–24533 (2019)
9. S. Hussain, M.S. Javed, S. Asim, A. Shaheen, A.J. Khan, Y. Abbas, S. Yun, Novel gravel-like NiMoO4 nanoparticles on carbon cloth for outstanding supercapacitor applications. Ceram. Int. **46**(5), 6406–6412 (2020)
10. V.S. Bhadram, D. Swain, R. Dhanya, M. Polentarutti, A. Sundaresan, C. Narayana, Effect of pressure on octahedral distortions in $RCrO_3$ (R = Lu, Tb, Gd, Eu, Sm): the role of R-ion size and its implications. Mater. Res. Express. **1**(2), 026111 (2014)
11. K. Yoshii, Magnetic properties of perovskite $GdCrO_3$. J. Solid State Chem. **159**(1), 204–208 (2001)

12. S. Kumar, I. Coondoo, M. Vasundhara, V.S. Puli, N. Panwar, Observation of magnetization reversal and magnetocaloric effect in manganese modified EuCrO3 orthochromites. Physica B **519**, 69–75 (2017). https://doi.org/10.1016/j.physb.2017.05.050

13. H. Li, Y.Z. Liu, L. Xie, Y.Y. Guo, Z.J. Ma, Y.T. Li, X.M. He, L.Q. Liu, H.G. Zhang, The spin-reorientation magnetic transitions in Ga-doped SmCrO3. Ceram. Int. **44**(15), 18913–18919 (2018). https://doi.org/10.1016/j.ceramint.2018.07.127

14. S. Hussain, X. Yang, M.K. Aslam, A. Shaheen, M.S. Javed, N. Aslam, G. Qiao, Robust TiN nanoparticles polysulfide anchor for Li–S storage and diffusion pathways using first principle calculations. Chem. Eng. J. **391**, 123595 (2020).

15. C.L. Chang, W. Wang, H. Ma, H. Huang, J.C. Liu, R.Z. Geng, Monte Carlo study of magnetic properties and magnetocaloric effect of an AFM/FM BiFeO3/Co bilayer. Commun. Theor. Phys. (2021)

16. G. Kadim, R. Masrour, A. Jabar, A comparative study between GGA, WC-GGA, TB-mBJ and GGA+ U approximations on magnetocaloric effect, electronic, optic and magnetic properties of BaMnS$_2$ compound: DFT calculations and Monte Carlo simulations. Phys. Scr. **96**(4), 045804 (2021)

17. M. Chakravorty, P.M. Chowdhury, Crystal field effect on the magneto caloric properties of the mixed spin-7/2 and spin-3/2 3D-ising ferrimagnet: a Monte Carlo simulation. J. Magn. Magn. Mater. **528**, 167818 (2021)

18. S. Yin, W. Zhong, C.J. Guild, J. Shi, S.L. Suib, L.F. Cótica, M. Jain, Effect of Gd substitution on the structural, magnetic, and magnetocaloric properties of HoCrO$_3$. J. Appl. Phys. **123**(5), 053904 (2018)

19. G. Kadim, R. Masrour, A. Jabar, Ferroelectric, quantum efficiency and photovoltaic properties in perovskite BiFeO$_3$ thin films: first principle calculations and Monte Carlo study. Int. J. Energy Res. 9961–9969 (2021)

20. E. Erbarut, Optical response functions of ZnS, ZnSe, ZnTe by the LOM method. Solid State Commun. **127**(7), 515–519 (2003)

21. G. Murtaza, B. Amin, S. Arif, M. Maqbool, I. Ahmad, A. Afaq, S. Nazir, M. Imran, M. Haneef, Structural, electronic and optical properties of Ca$_x$Cd$_{1-x}$O and its conversion from semimetal to wide bandgap semiconductor. Comput. Mater. Sci. **58**, 71–76 (2012)

22. Z.J. Li, Z.L. Hou, W.L. Song, X.D. Liu, W.Q. Cao, X.H. Shao, M.S. Cao, Unusual continuous dual absorption peaks in Ca-doped BiFeO$_3$ nanostructures for broadened microwave absorption. Nanoscale **8**(19), 10415–10424 (2016)

23. H. Lashgari, M.R. Abolhassani, A. Boochani, E. Sartipi, R. Taghavi-Mendi, A. Ghaderi, Ab initio study of electronic, magnetic, elastic and optical properties of full Heusler Co$_2$MnSb. Indian J. Phys. **90**(8), 909–916 (2016)

24. R. Khenata, A. Bouhemadou, M. Sahnoun, A.H. Reshak, H. Baltache, M. Rabah, Elastic, electronic and optical properties of ZnS, ZnSe and ZnTe under pressure. Comput. Mater. Sci. **38**(1), 29–38 (2006)

25. R.A. Da Silva, R.N. Saxena, A.W. Carbonari, G.A. Cabrera-Pasca, Investigation of hyperfine interactions in GdCrO$_3$ perovskite oxide using PAC spectroscopy. Hyperfine Interact **197**(1), 53–58 (2010)

# Chapter 5
# Magnetic Properties and Magnetocaloric in Double Perovskite Oxides

**Abstract** We have studied the magnetic properties and magnetocaloric effect in $Sr_2FeMoO_6$ compound by using the Monte Carlo study. Thermal magnetization, M and dM/dT of $Sr_2FeMoO_6$ are investigated. The transition temperature and lock-in-transition temperatures are deduced. The temperature dependence of the magnetic entropy and of the adiabatic temperature for a several magnetic fields have been also obtained. The field dependence of relative cooling power and magnetic hysteresis cycle of $Sr_2FeMoO_6$ have been determined for a several magnetic field and temperatures. This system might be a promising base for developing the new kinds of magnetic refrigerant as working materials in magnetic refrigeration technology.

## 5.1 Introduction

The magnetic and resistive properties of the double perovskite $Sr_2FeMoO_6$ are excellent for spintronic and magnetoresitive applications [1]. Kobayashi et al. [2], reported a large magnetoresistance effect with a high magnetic transition temperature of about 418 K in $Sr_2FeMoO_6$, a material belonging to the class of double perovskites ($ABB'O_6$), where the alkaline-earth ion A is Sr and transition-metal ions B and B' are Fe and Mo arranged in a rock-salt structure, respectively [3]. Magnetism testing results show that the sample $Sr_2FeMoO_6$ is ferromagnetic with the magnetic transition temperature of about 380K [4]. The double perovskite crystal structure consists of two interpenetrating face-centered-cubic sublattices. An ordered arrangement of $Fe^{3+}$ ($3d^5$) and S = 5/2 magnetic moments antiferromagnetically coupled to the $Mo^{5+}$ and S = 1/2 moments gives a total saturation magnetic moment of $4\mu_B$ at low temperature [5, 6]. The Fe ions present an ionic $Fe^{3+}$ ($3d^5$) valence, whereas the Mo ions are in a strongly covalent $Mo^{5+}$ ($4d^1$) state [7]. In other hand the ordered double-perovskite $Sr_2FeMoO_6$ possesses remarkable room-temperature low-field colossal magneto-resistivity and transport properties which are related, at least in part, to combined structural and magnetic instabilities that are responsible for a cubic-tetragonal phase transition near 420K [8]. The latter condition is proving problematic: the substitution of other 3d metals for $Fe^{3+}$ can give rise to the formation

© The Author(s), under exclusive license to Springer Nature Switzerland AG 2024
R. Masrour, *Magnetoelectronic, Optical, and Thermoelectric Properties of Perovskite Materials*, SpringerBriefs in Materials,
https://doi.org/10.1007/978-3-031-48967-9_5

of an $M^{2+}/Mo^{6+}$ ion pair [9–11]. Moreover, the magnetic properties of $Sr_2FeMoO_6$ ceramics obtained by sol–gel and solid-state reaction methods and sintered by the classical method were compared with those of $Sr_2FeMoO_6$ ceramics obtained by the same two methods, but sintered by the spark plasma sintering technique [12] and neutron diffraction and magnetic susceptibility has been also used to study the crystalline and magnetic structures of $Sr_2FeMoO_6$ by Refs. [13, 14]. The density-functional theory, the effect of biaxial mechanical strain on the magnetic properties of double perovskite oxide $Sr_2FeMoO_6$ has been studied [15]. The saturation magnetization originating mainly from the Fe moments is correlated with the amount of Mo magnetic moments observed by nuclear magnetic resonance measurements in $Sr_2FeMoO_6$ [16, 17]. The paper is organized as follow. Section 5.2 present the Model, Sects. 5.2 and 5.3 present the Monte Carlo simulations and results and discussion and finally we have given in Sect. 5.4 the conclusions.

## 5.2  Ising Model and Monte Carlo Simulations

In this section, for explanation of MC algorithm, a following Hamiltonian described the $Sr_2FeMoO_6$ compound using Ising model includes nearest neighbors interactions and external magnetic field h and crystal filed $\Delta$ is given by:

$$H = -J_{FeFe1} \sum_{<i,j>} S_i S_j - J_{FeFe2} \sum_{<<i,k>>} S_i S_k - J_{FeMo1} \sum_{<i,l>} S_i \sigma_l - J_{FeMo2} \sum_{<<i,m>>} S_i \sigma_m$$
$$- J_{MoMo1} \sum_{<m,n>} \sigma_m \sigma_n - J_{MoMo2} \sum_{<<m,p>>} \sigma_m \sigma_p - h \left( \sum_i S_i + \sum_l \sigma_l \right)$$

where $S_i(Fe^{3+}) = 5/2$ and $\sigma_i(Mo^{3+}) = 1/2$ denotes the spin states of $i$-th cell. The ($\langle i, j \rangle$, $\langle i, l \rangle$ and $\langle m, n \rangle$) and ($\langle\langle i, k \rangle\rangle$, $\langle\langle i, m \rangle\rangle$ and $\langle\langle m, p \rangle\rangle$) stand for the first and second nearest neighbor sites (($i$ and $j$), ($i$ and $l$) and ($m$ and $n$)) and (($i$ and $k$), ($i,m$) and ($m,p$)), respectively. The ($J_{FeFe1}$, $J_{FeMo1}$ and $J_{MoMo1}$) and ($J_{FeFe2}$, $J_{FeMo2}$ and $J_{MoMo2}$) are the first and second exchange interactions between ($Fe$-$Fe_1$, $Fe$-$Mo$ and $MoMo_1$) and ($Fe$-$Fe_2$, $Fe$-$Mo_2$ and $Mo$-$Mo_2$) such as shown in Fig. 5.1.

The $Sr_2FeMoO_6$ compound is assumed to reside in the unit cells and the system consists of the total number of spins $N = N_S + N_\sigma$ with $N_S = 2331$ and $N_\sigma = 2310$. The Monte Carlo simulation is applied to simulate the Hamiltonian given by Eq. (1). In this section, we have used the cyclic boundary conditions on the lattice of $Sr_2FeMoO_6$ compound. The Monte Carlo update was performed by choosing random spins and then flipped (from current state $S_i(\sigma_i)$ to opposite state $-S_i(-\sigma_i)$) with Boltzmann based probability. In general, this can be done using the conventional Metropolis algorithm [18] i.e. $P_{Metro} = \exp(-\Delta E/k_B T)$ where $\Delta E$ is the energy difference between the before and the after flip. However, since the actual Metropolis suffers from large correlation time especially close to the critical point [19].

**Fig. 5.1** Crystal structure of double $Sr_2FeMoO_6$ perovskite

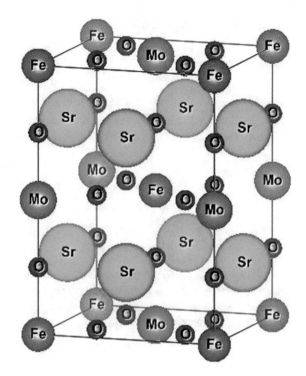

Our program calculates the following parameters, namely:
The internal energy per site $E$ is of $Sr_2FeMoO_6$ compound,

$$E = \frac{1}{N}\langle H \rangle$$

The total magnetization of $Sr_2FeMoO_6$ compound is:

$$M = \left\langle \frac{1}{N}\sum_i S_i \right\rangle$$

The magnetic specific heat of $Sr_2FeMoO_6$ compound is given by:

$$\frac{C_m}{\beta} = \frac{k_B}{N}\left(\langle E^2 \rangle - \langle E \rangle^2\right)$$

where $\beta = \frac{1}{k_BT}$, T denotes the absolute temperature.
The thermodynamic Maxwell relation is given by:

$$\left(\frac{\partial S}{\partial h}\right)_T = \left(\frac{\partial M}{\partial T}\right)_h$$

The magnetic entropy of a material is given by;

$$S(T, h) = \int\limits_{0}^{T} \frac{C_m}{T} dT$$

with $C_m$ is the specific heat for the magnetic structure.

The magnetic entropy changes between $h$ different to zero and $h = 0$ is:

$$\Delta S_m(T, h) = S_m(T, h) - S_m(T, 0) = \int\limits_{0}^{h_{max}} \left( \frac{\partial M}{\partial T} \right)_{h_i} dh$$

$$= \sum_i \left( \frac{\partial M}{\partial T} \right)_{h_i} \Delta h_i$$

$S_m$ $(T, h)$ and $S_m$ $(T, 0)$ are the total entropy in presence and absence of magnetic field, respectively and $h_{max}$ is the maximum applied external magnetic field. $\left( \frac{\partial M}{\partial T} \right)_{h_i}$ is the thermal magnetization for a fixed magnetic field $h_i$.

The adiabatic temperature from Eq. (6) is given:

$$\Delta T_{ad} = -T \frac{\Delta S_m}{C_{p,h}}$$

The defined parameter of relative cooling power (RCP) described as an area under the dependence of $\Delta S_m(T)$ on temperature, is a compromise between the magnitude of the magnetic entropy change and the width of the peak. The expression of relative cooling power RCP is:

$$RCP = \int\limits_{T_c}^{T_h} \Delta S_m(T) dT$$

where $T_c$ and $T_h$ are the cold and the hot temperatures corresponding to both ends of the half-maximum value of $\Delta S_m^{max}$, respectively.

The excitation magnetic field B is given by: $B = h + 4\pi M$.

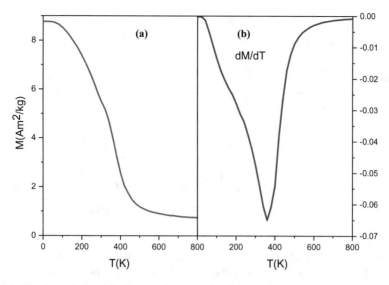

**Fig. 5.2** The thermal magnetization **a** and dM/dT of $Sr_2FeMoO_6$ compound for h = 4

## 5.3 Results and Discussion of Magnetic Properties and Magnetocaloric in Double Perovskite Oxides

The Magnetic properties and magnetocaloric effect of the $Sr_2FeMoO_6$ compound are studied using Monte Carlo simulations. We have presented in Fig. 5.2a and b, the thermal magnetization and dM/dT, respectively for h = 4 T. The Curie temperature is obtained $T_C = 460$ K and is comparable with $T_C = 358$ K that given in reversible magnetocaloric response in $Sr_2FeMoO_6$ double perovskite study [20] and is lower to those 380K and 420 K reported in Refs. [4] and [8], respectively. Therefore, the low value of $T_C$ observed in the present work is due besides anti-site defect and oxygen vacancy and the presence of nonmagnetic $SrMoO_4$ secondary phase which increases the Fe content in the $Sr_2FeMoO_6$ sample leads to antiferromagnetic Fe–O–Fe interactions and produce the anti-site defects. The temperature dependence of the magnetic entropy change of $Sr_2FeMoO_6$ compound has been shown in Fig. 5.3, for several external magnetic field h = 2, 4, 8 and 12. We have also deduced the Curie temperature $T_C = 360$ K from the maximum of $-\Delta S$ (T, h). $- \Delta S^{max}$ shows linear dependence of $h^{2/3}$ indicating a second-order magnetic phase transition. These results are comparable with those given in field dependence of the magnetocaloric effect in materials with a second order phase transition [21]. Figure 5.4a, shows the temperature dependence of the adiabatic temperature change of $Sr_2FeMoO_6$ compound for several external magnetic field h = 2, 4, 8 and 12 T. The Curie temperature is a obtained and is comparable with that given in Figs. 5.2 and 5.3. In addition to this transition, all the samples are also found to exhibit another transition, known as the lock-in-transition ($T_{lock}$). The obtained results are comparable with those given in

**Fig. 5.3** The temperature dependence of the magnetic entropy changes and the Inset shows the plot of $-\Delta S^{max}$ versus $h^{2/3}$ of $Sr_2FeMoO_6$ compound for several external magnetic field h = 2, 4, 8 and 12 T

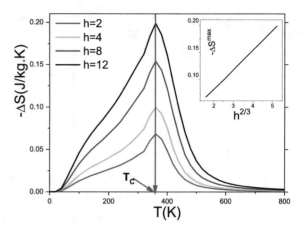

previous work [22–24]. The obtained values of adiabatic temperature $\Delta T_{ad}$ for h = 2, 4, 8 and 12 are 0.012, 0.018, 0.031 and 0.043 J/kg.K, respectively. The $\Delta T_{ad}$ increases with increasing the values of external magnetic field. Fig. 5.4b presents, the field dependence of relative cooling power (RCP) of $Sr_2FeMoO_6$ compound. The obtained values of RCP increase with increasing the external magnetic field. This result is comparable with that given in previous work [25]. The value of RCP for h = 12 is 18.5 J/kg for T = 800 K. We have also presented in Fig. 5.5a, the Magnetic hysteresis cycle of $Sr_2FeMoO_6$ compound for several temperatures T = 200, 300, 360 and 420 K. The magnetic coercive field decreases with increasing the temperatures values. The superparamagnetic behavior is observed around the transition temperature $T_C$. This result is similar to that obtained in Ref. [26]. Finally, we have given in Fig. 5.5b, the magnetic excitation B versus the external magnetic field h for different temperatures T = 0.1, 0.5, 1 and 2 K. The superparamagnetic behavior is observed at low temperature.

## 5.4  Conclusions

The magnetic properties and magnetocaloric effect in $Sr_2FeMoO_6$ compound have been investigated using Monte Carlo simulations. The Curie and lock-in-transition temperatures have been obtained and are comparable with those given by experiment results. The magnetic entropy change and the adiabatic temperature change show a huge jump at the ferromagnetic transition, which is usually observed when the transition is second order like. The $\Delta S_{max}$ increases with increasing the external magnetic field. The magnetic hysteresis cycles are found for different temperatures. The coercive fields decrease with increasing the temperatures values. The large $\Delta S$

**Fig. 5.4** **a** The temperature dependence of the adiabatic temperature change of $Sr_2FeMoO_6$ compound for several external magnetic field h = 2, 4, 8 and 12 T. **b** The field dependence of relative cooling power (RCP) of $Sr_2FeMoO_6$ compound

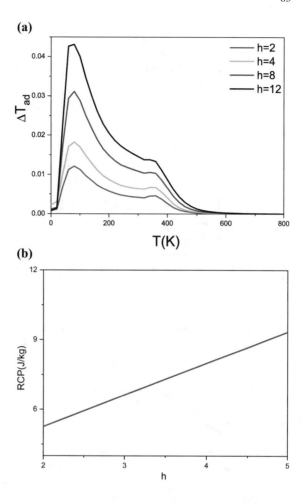

together with the absence of hysteresis make $Sr_2FeMoO_6$ a potential candidate material ferromagnetic refrigeration. The system shows the superparamagnetic behavior around the Curie temperature and at low temperatures values.

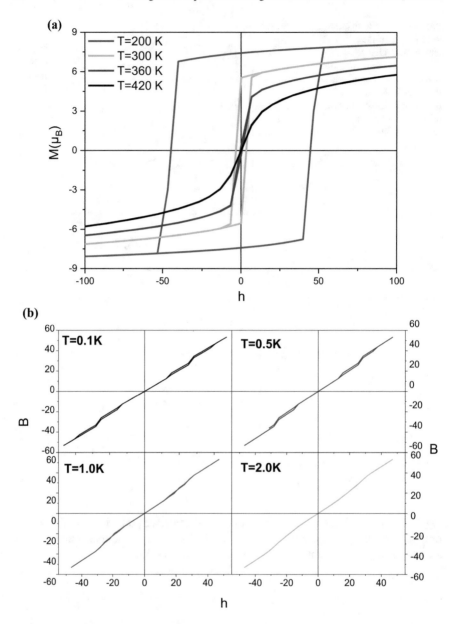

**Fig. 5.5** **a** The Magnetic hysteresis cycle of $Sr_2FeMoO_6$ compound for several temperatures T = 200, 300, 360 and 420 K. **b** The magnetic excitation B versus the external magnetic field for different temperatures T = 0.1, 0.5, 1 and 2 K

# References

1. M. Saloaro, M. Hoffmann, W.A. Adeagbo, S. Granroth, H. Deniz, H. Palonen, H. Huhtinen, S. Majumdar, P. Laukkanen, W. Hergert, A. Ernst, P. Paturi, ACS Appl. Mater. Interfaces **8**, 20440–20447 (2016)
2. K.-I. Kobayashi, T. Kimura, H. Sawada, K. Terakura, Y. Tokura, Nature **395**, 677 (1998)
3. T. Saha-Dasgupta, D.D. Sarma, Phys. Rev. B **64**, 064408 (2001)
4. Y. Zhai, J. Qiao, G. Huo, S. Han, J. Magn. Magn. Mater. **324**, 2006–2010 (2012)
5. F. S. Galasso, Structure, properties, and preparation of Perovskite-type compounds (Pergamon, Oxford, 1969), Chap. 2
6. C.M. Hurd, *The Hall Effect in Metals and Alloys* (Plenum, New York, 1972)
7. H.P. Martins, F. Prado, A. Caneiro, F.C. Vicentin, D.S. Chaves, R.J.O. Mossanek, M. Abbate, J. Alloys Compd. **640**, 511–516 (2015)
8. D. Yang, R.J. Harrison, J.A. Schiemer, G.I. Lampronti, X. Liu, F. Zhang, H. Ding, Yan'gai Liu, and Michael A. Carpenter. Phys. Rev. B **93**, 024101 (2016)
9. S. Nomura, T. Nakagawa, J. Phys. Soc. Jpn. **21**, 1068 (1966)
10. M.C. Viola, M.J. Martinez-Lope, J.A. Alonso, P. Velasco, J.L. Martinez, J.C. Pedregosa, R.E. Carbonio, M.T. Fernandez-Diaz, Chem. Mater. **14**, 812 (2002)
11. A.K. Azad, S.-G. Eriksson, S.A. Ivvanov, R. Mathieu, S.P. Svedlindh, J. Eriksen, H. Rundlof, J. Alloys Compd. **364**, 77 (2004)
12. M. Cernea, F. Vasiliu, C. Bartha, C. Plapcianu, I. Mercioniu, Ceram. Int. **40**, 11601–11609 (2014)
13. M.K. Chung, P.J. Huang, W.-H. Li, C.C. Yang, T.S. Chan, R.S. Liu, S.Y. Wud, J.W. Lynn, Physica B **385–386**, 418–420 (2006)
14. D. Sanchez, J.A. Alonso, M. Garcıa-Hernandez, M.J. Martınez-Lope, J.L. Martınez, Phys. Rev. B **65**, 104426 (2002)
15. Ruifeng Lu, Haiping Wu, Yan Qian, Erjun Kan, Yuzhen Liu, Weishi Tan, Chuanyun Xiao, Kaiming Deng. Solid. State. Commun. **191**, 70–75 (2014)
16. S. Colis, G. Pourroy, P. Panissod, C. M!eny, A. Dinia. J. Magn. Magn. Mater. 272–276 (2004) 2018–2020
17. K. Ohno, H. Kato, T. Nishioka, M. Matsumura, J. Magn. Magn. Mater. **310**, e666–e668 (2007)
18. N. Metropolis, A.W. Rosenbluth, M.N. Rosenbluth, A.H. Teller, E. Teller, J. Chem. Phys. **21**, 1087 (1953)
19. M.E.J. Newman, G.T. Barkema, *Monte carlo methods in statistical physics* (Oxf. Univ. Press., Oxford, 1999)
20. M.S. Anwar, Imad Hussain. Bon Heun Koo. Mater. Lett. **181**, 56–58 (2016)
21. V. Franco, J.S. Blázquez, A. Conde, Appl. Phys. Lett. **89**, 222512 (2006)
22. N. Pavan Kumar, P. Venugopal Reddy. Mater. Lett. 122 (2014) 292
23. K. Noda, S. Nakamura, J. Nagayama, H. Kuwahara, J. Appl. Phys. **97**, 10C103 (2005)
24. S. Takai, T. Matsumura, H. Tanida, M. Sera, Phys. Rev. B **92**, 174427 (2015)
25. R. Cherif, E.K. Hlil, M. Ellouze, F. Elhalouani, S. Obbade, J. Solid State Chem. **215**, 271 (2014)
26. Y.M. Hao, S.Y. Lou, S.M. Zhou, R.J. Yuan, G.Y. Zhu, N. Li, Nanoscale Res. Lett. **7**, 100 (2012)

# Chapter 6
# Magnetocaloric and Magnetic Properties of Bilayer Manganite

**Abstract** The structure of bilayer manganite $La_2SrMn_2O_7$ has been investigated using the Monte Carlo simulations. Thermal magnetization of bilayer manganite system was investigated. We have given the dM/dT as a function of temperatures to find the transitions temperatures. The magnetic transition from ferromagnetic to paramagnetic is found. The second-order phase transition is found at the transition temperature. The temperature dependence of the magnetic entropy changes of temperatures of $La_2SrMn_2O_7$ bilayer manganite for a several external magnetic fields was found. The field dependence of relative cooling power of $La_2SrMn_2O_7$ bilayer manganite is given. Finally, the magnetic hysteresis cycle of bilayer manganite system was given. The superparamagnetic behavior has been found around the transition temperature.

## 6.1 Introduction

Experimentally, the $LaSr_2Mn_2O_7$ has been investigated by Refs. [1–3]. Double-layered manganites with the general formula $LaSr_2Mn_2O_7$ have attracted extensive interest due to discovery of colossal magnetoresistance, charge ordering, and different magnetic phases [4, 5]. One potential application of these materials relates also to porous cathodes of solid oxide fuel cells, the alternative power generation systems enabling to increase energy conversion efficiency [6]. A detailed study of the electronic structure and magnetic configurations of the 50% hole-doped double-layered manganite $LaSr_2Mn_2O_7$ was studied [7]. In previous study, in order to investigate the effect of particle size, we prepared a number of $LaSr_2Mn_2O_7$ ceramic samples using the sol–gel method, and the effects of particle size on the electrical properties of this compound were reported [8]. the magnetic and transport properties in the mixed-valence manmites can be understood in terms of double-exchange interaction mehanism [9]. The electrical transition temperature and magnetoresistance prediction of $LaSr_2Mn_2O_7$ bilayered manganite have been studied experimentally [10]. In previous works [11, 12], we have used the the Monte Carlo to study the Surface behavior of magnetic phase transitions and the magnetocaloric properties of perovskites ferromagnetic thin films. Magnetoresistance and magnetization were

R. Masrour, *Magnetoelectronic, Optical, and Thermoelectric Properties of Perovskite Materials*, SpringerBriefs in Materials, https://doi.org/10.1007/978-3-031-48967-9_6
87

studied for $LaSr_2Mn_2O_7$ [13]. In addition, in the $LaSr_2Mn_2O_7$ with a grain size of 200 nm, enhancement of the magnetic properties, which is accompanied with the formation of ferromagnetic phase on the surface of grain or particle, is observed [14].

The chapter book is organized as follows: In the Sect. 6.2 we define the model and theory and in Sect. 6.3 the used method. Resulting and discussed of magnetic properties of LSMO bilayer manganite are presented in Sect. 6.4. The final Sect. 6.5 is devoted to conclusions.

## 6.2   Ising Model and Monte Carlo Simulations

The LSMO system crystallizes in tetragonal with parameters lattices a = 3.868 and c = 20.205 Å with space group I4/mmm [15]. We have used the positional coordinates of La, Sr, Mn and O in LSMO given in Ref. [15] and their occupation numbers and symmetry (see Table. 6.1) to present the structure of our system such as given in Fig. 6.1 in the form of bilayer. The Hamiltonian described our compounds using Ising model includes nearest neighbors interactions and external magnetic field $h$ is used:

$$H = -J_1 \sum_{<i,j>} S_i S_j - J_2 \sum_{<<i,k>>} S_i S_k - J_3 \sum_{<<<i,m>>>} S_i S_m$$
$$- J_4 \sum_{<<<<i,n>>>>} S_i S_n - h \sum_i S_i$$

The Full-Potential Linear Augmented Plane Wave as implemented in WIEN2k code [16] to calculate the values of first, second, third, *and* fourth nearest neighbors. *The obtained values* of $J_1 = 13.6$, $J_2 = 11.7$, $J_3 = 10.3$ and $J_4 = 9.1$ K, respectively. $(La^{3+})_2Sr^{2+}(Mn^{3+})_2(O^{2-})_7$ consists only $Mn^{3+}$ with moment of spin is $S = 2$ such as given in Ref. [17]. The exchange interactions $J_1$, $J_2$, $J_3$ and $J_4$ are represented in Fig. 6.1.

**Table 6.1** Positional coordinates of La, Sr, Mn, O [15] and their occupation numbers and symmetry in $La_2SrMn_2O_7$

| Atoms | x | y | z | Occupation numbers | Symmetry |
|-------|---|---|---|--------------------|----------|
| La/Sr(1) | 0 | 0 | 0.5 | 1.000 | 4/mmm |
| La/Sr(2) | 0 | 0 | 0.312 | 1.000 | 4mm |
| Mn | 0 | 0 | 0.097 | 1.000 | 4mm |
| O | 0 | 0 | 0 | 1.000 | 4/mmm |
| O | 0 | 0 | 0.189 | 1.000 | 4mm |
| O | 0 | 0.5 | 0.093 | 1.000 | 2mm |

**Fig. 6.1** Structure of
La$_2$SrMn$_2$O$_7$ bilayer
manganite with: The red
sphere is atom of oxygen; the
green sphere is atom of
Lanthanum or Strontium and
the purple sphere is atom of
Magnesium

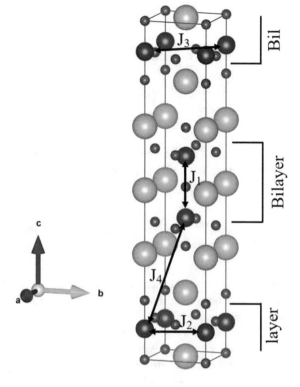

The La$_2$SrMn$_2$O$_7$ bilayer manganite system with L = 20 is assumed to reside in
the unit cells and the system consists of the total number of spins N = 6260 with the
number of bilayers is 20. We have used the Monte Carlo simulation to simulate the
Hamiltonian. The cyclic boundary conditions on the system were imposed in plane (a,
b) and free system in c axe. Monte Carlo update was performed by choosing random
spins and then flipped with Boltzmann based probability by using the Metropolis
algorithm [18].

Our program calculates the following parameters, namely.

The internal energy per site $E$ is of LSMO bilayer manganite is:

$$E = \frac{1}{N}\langle H \rangle$$

The magnetization of LSMO bilayer manganite is given by:

$$M = \left\langle \frac{1}{N} \sum_{i(i \in bilayers)} S_i \right\rangle$$

The magnetic entropy changes is:

$$\Delta S_m(T, h) = \sum_i \left(\frac{\partial M}{\partial T}\right)_{hi} \Delta hi$$

The expression of relative cooling power RCP is:

$$RCP = \int_{T_c}^{T_h} \Delta S_m(T) dT$$

where $T_c$ and $T_h$ are the cold and the hot temperatures corresponding to both ends of the half-maximum value of $\Delta S_m^{max}$, respectively.

## 6.3   Magnetic and Magnetocaloric Properties of Bilayer Manganite System

The magnetic and magnetocaloric properties of $La_2SrMn_2O_7$ bilayer manganite system have been investigated using Monte Carlo. We have presented in Fig. 6.2, thermal magnetization of LSMO bilayer manganite for a several external magnetic fields h = 1–5 T. For a fixed value of temperature, the magnetization increases with increasing the external magnetic field. The magnetization decreases with increasing the temperatures values for each value of external magnetic field.

Figure 6.3 illustrates the variation of dM/dT with temperatures of LSMO bilayer manganite with a several external magnetic fields h = 1 to 5 T. This curve exhibited a sharp magnetic transition from paramagetic to ferromagnetic such as given in previous work [15]. The transition temperature, which is defined as the minimum

**Fig. 6.2** Thermal magnetization of $La_2SrMn_2O_7$ bilayer manganite with a several external magnetic fields h = 1, 3 and 5 T

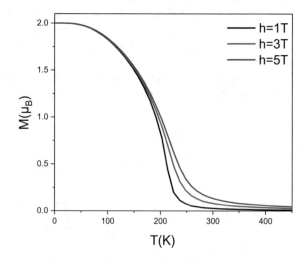

**Fig. 6.3**  dM/dT as a function of temperatures of La$_2$SrMn$_2$O$_7$ bilayer manganite with a several external magnetic fields h = 1, 3 and 5 T

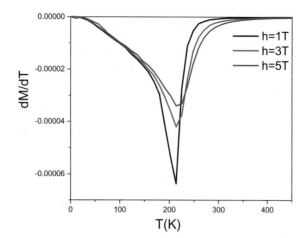

in dM/dT, was determined as 213.76 K. The obtained value of T$_C$ is comparable with that obtained by was measured by Faraday technique T$_C$ = 210 K [15]. The second-order phase transition is obtained at the transition temperature.

Figure 6.4 shows, the temperature dependence of the magnetic entropy change of temperatures of LSMO bilayer manganite for a several external magnetic fields h = 1, 3 and 5 T. From the Fig. 6.4, we see that the magnetic entropy change depends on both external magnetic field and temperatures, where ($-\Delta$S) increased and reached a maximum value when the temperature approached the transition temperature T$_C$(K) = 213.76 K. The values of maximum of magnetic change entropy are: $\Delta S_m^{max}$ = 0.39, 0.68, 1.03 and 1.23 J/kg · K for h = 1, 3 and 5 T, respectively. The value of $\Delta S_m^{max}$ = 1.23J/kg · K is near to that obtained by Ref. [19] for a weak magnetic field. In addition, a large magnetocaloric effect has also been found in this system. For example, a notable magnetic entropy change of 16.8 J/kg · K under a magnetic field change of 5.0 T was reported for La$_{1.4}$Sr$_{1.6}$Mn$_2$O$_7$ [20]. This value is much larger than that determined for pure Gd (10.2 J/kg · K) under the same magnetic field change [21], which is considered as a reference for magnetocaloric materials and Gd$_5$Ge$_2$Si$_2$ or MnFeP$_{1-x}$As$_x$ compounds (18 J/kg · K) at the magnetic field 5T [22].

Figure 6.5 shows the field dependence of relative cooling power of LSMO bilayer manganite for T = 450 K. The field dependence of relative cooling power increases with increasing the external magnetic field for a fixed temperature value. The maximum value of RCP for h = 5T is 11 J/kg.

Figure 6.6 gives the magnetic hysteresis cycle of LSMO bilayer manganite for a several temperatures T = 160, 190, 213 and 220 K. The magnetic coercive field and remanant magnetization decrease with increasing the temperatures value and becomes equal to zero around the transition temperature. The superparamagnetic behavior is obtained for T = 213 and 220 K.

**Fig. 6.4** Temperature
dependence of the magnetic
entropy change of
temperatures of
La$_2$SrMn$_2$O$_7$ bilayer
manganite for a several
external magnetic fields h =
1, 3 and 5 T

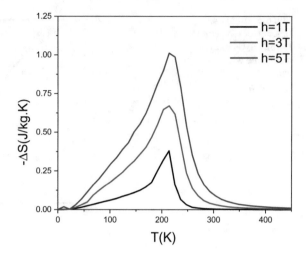

**Fig. 6.5** The field
dependence of relative
cooling power of
La$_2$SrMn$_2$O$_7$ bilayer
manganite for T = 450 K

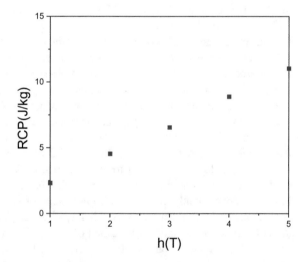

## 6.4  Conclusions

In this chapter, the magnetic and magnetocaloric properties of La$_2$SrMn$_2$O$_7$ by using
Monte Carlo simulation. The paramagnetic—ferromagnetic phase transition has been
obtained around the transition temperature. The phase transition was confirmed as
a second order by analyzing the isothermal magnetic behavior. The maximum of
magnetic entropy changes increases with increasing the external magnetic field and
it is situated at the transition temperature. The relative cooling found increases with
increasing the external magnetic field. The magnetic coercive field and remanant

**Fig. 6.6** The Magnetic hysteresis cycle of $La_2SrMn_2O_7$ bilayer manganite for a several temperatures T = 160, 190, 213 and 220 K

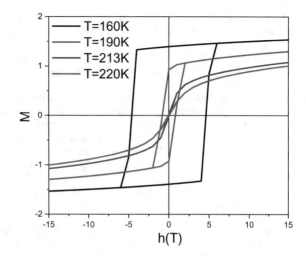

magnetization decrease with increasing the temperatures value. Superparamagnetic behavior is obtained around the transition temperature.

# References

1. J.F. MacChesney, J.F. Potter, R.C. Sherwood, Chemical and magnetic study of layered strontium lanthanum manganate structures. J. Appl. Phys. **40**, 1243–1245 (1969)
2. R. Ram, P. Ganguly, C.N.R. Rao, Magnetic properties of quasi-two-dimensional $La_{1-x}Sr_{1+x}MnO_4$ and the evolution of itinerant electron ferromagnetism in the SrO. $(La_{1-x}Sr_xMnO_3)_n$ system. J. Solid State Chem. **70**, 82–87 (1987)
3. R.S. Kumar, D. Prabhakaran, A. Boothroyd et al., Structural studies of $La_{2-2x}Sr_{1+2x}Mn_2O_7$ bilayer manganites. J. Phys. Chem. Solids **67**, 2046–2050 (2006)
4. Y. Moritomo, A. Asamitsu, H. Kuwahara et al., Giant magnetoresistance of manganese oxides with a layered perovskite structure. Nature **380**, 141–144 (1996)
5. T. Kimura, Y. Tokura, Layered magnetic manganites—annual reviews. Ann. Rev. Mater. Sci. **30**, 451–474 (2000)
6. A.A. Yaremchenko, D.O. Bannikov, A.V. Kovalevsky et al., High-temperature transport properties, thermal expansion and cathodic performance of Ni-substituted $LaSr_2Mn_2O_{7-\delta}$. J. Solid State Chem. **181**, 3024–3032 (2008)
7. J.E. Medvedeva, V.I. Anisimov, M.A. Korotin, O.N. Mryasov, A.J. Freeman, Coulomb correlation and magnetic ordering in double-layered manganites: $LaSr_2Mn_2O_7$. J. Magn. Magn. Mater. **237**, 47–54 (2001)
8. M.H. Ehsani, M.E. Ghazi, P. Kameli, Tunable magnetic and magnetocaloric properties of $La_{0.6}Sr_{0.4}MnO_3$ nanoparticles. J. Mater. Sci. **47**, 5815–5825 (2012)
9. C. Zener, Interaction between the d-shells in the transition metals. II. Ferromagnetic comyountls of manganese with perovskite structure. Phys. Rev. **82**, 403–405 (1951)
10. M.H. Ehsani, M.J. Mehrabad, The electrical transition temperature and magnetoresistance prediction of $LaSr_2Mn_2O_7$ bilayered manganite. J. King. Saud. Univ. Sci. **I**(30), 339–344 (2018)
11. R. Masrour, A. Jabar, Surface behavior of magnetic phase transitions: a Monte Carlo study. Appl. Surf. Sci. **432**, 78–84 (2018)

12. A.S. Erchidi Elyacoubi, R. Masrour, A. Jabar, Surface effects on the magnetocaloric properties of perovskites ferromagnetic thin films: a Monte Carlo study. Appl. Surf. Sci. **459**, 537–543 (2018)
13. T. Hayashi, N. Miura, M. Tokunaga et al., Magnetic properties and CMR effect in layer type manganite $LaSr_2Mn_2O_7$ under high magnetic fields. J. Phys.: Condens. Matter **10**, 11525–11529 (1998)
14. M.H. Ehsani, M.E. Ghazi, P. Kameli et al., DC magnetization studies of nano- and micro-particles of bilayered manganite $LaSr_2Mn_2O_7$. J. Alloys. Compnds. **586**, 261–266 (2014)
15. I.B. Sharma, D. Singh, Preparation and study of structural, electrical and magnetic properties of $La_2SrFe_2O_7$ and LSMO. Proc. Indian Acad. Sci. **10**, 189–196 (1995)
16. P. Blaha, K. Schwarz, G.K.H. Madsen et al., WIEN2k, an augmented plane wave plus local orbitals program for calculating crystal properties. Vienna University of Technology, Vienna, Austria (2001)
17. H. Martinho, C. Rettori, D.L. Huber et al., Low-energy spin-wave excitations in the bilayered magnetic manganite $La_{2-2x}Sr_{1+2x}Mn_2O_7 (0.30 \leq x \leq 0.50)$. Phys. Rev. B **67**, 214428–214435 (2003)
18. N. Metropolis, A.W. Rosenbluth, M.N. Rosenbluth et al., Equation of state calculations by fast computing machines. J. Chem. Phys. **21**, 1087–1092 (1953)
19. Y.E. Yang, Y. Xie, L. Xu et al., Structural, magnetic, and magnetocaloric properties of bilayer manganite $La_{1.38}Sr_{1.62}Mn_2O_7$. J. Phys. Chem. **115**, 311–316 (2018)
20. H. Zhu, H. Song, Y.H. Zhang, Magnetocaloric effect in layered perovskite manganese oxide $La_{1.4}Sr_{1.6}Mn_2O_7$. Appl. Phys. Lett. **81**, 3416–3418 (2002)
21. V.K. Pecharsky, K.A. Gschneidner, A.O. Tsokol, Recent developments in magnetocaloric materials. Rep. Prog. Phys. **68**, 1479 (2005)
22. O. Tegus, E. Bruck, K.H.J. Buschow et al., Nature **415**, 150–152 (2002)

# Chapter 7
# Magnetocaloric Properties of Surface Effects in Perovskites Ferromagnetic Thin Films

**Abstract** The Monte Carlo simulations were used to study the surface effects on the magnetocaloric properties of ferromagnetic perovskite thin films. The thermal magnetization, transition temperatures, magnetic entropy change, relative cooling power, and magnetic hysteresis cycle temperature as a function of film thickness and exchange coupling were found. of surface. The reduced critical temperature of perovskite ferromagnetic thin films is studied as a function of the film thickness and the exchange interactions in the bulk, in the surface and between the surfaces. We have shown that the maximum entropy change in thin-film systems can be observed at temperatures well below the magnetic phase transition temperature. Maximum entropy changes increase with increasing external magnetic field. The relative cooling power increases with increasing external magnetic field and depends on the film thickness. The relative cooling power decreases with increasing surface exchange coupling value. The magnetic coercive field decreases with increasing temperature values when the surface exchange coupling is lower than the volume exchange coupling. The magnetic coercive field of perovskite ferromagnetic thin films is investigated as a function of reduced surface exchange coupling.

## 7.1 Introduction

Many studies on manganite films have primarily concentrated on epitaxial thin films grown on single crystal substrates [1]. The sol–gel technique has emerged as a dependable method for synthesizing a broad range of materials [2], including thick layers suitable for competitive magnetic sensors with low cost and low field sensitivity. Simultaneously, magnetic refrigeration based on the magnetocaloric effect is emerging as a promising technology poised to replace conventional vapor compression refrigeration due to its superior energy efficiency, compact size, and reduced environmental impact [3–8]. The magnetocaloric effect is principally determined by two critical parameters: the magnetic entropy change and the adiabatic temperature change in response to a magnetic field variation, both of which are pivotal for achieving high magnetic refrigeration efficiency [9]. Consequently, recent research

© The Author(s), under exclusive license to Springer Nature Switzerland AG 2024
R. Masrour, *Magnetoelectronic, Optical, and Thermoelectric Properties of Perovskite Materials*, SpringerBriefs in Materials,
https://doi.org/10.1007/978-3-031-48967-9_7

efforts have been heavily invested in the development of novel magnetocaloric materials at large [10–12].

The MnCoGe compound exhibits excellent magnetic properties, facilitating the measurement of perpendicular and parallel magnetocrystalline anisotropy constants up to 100 K. The Chappert model was employed to fit the ferromagnetic resonance measurements [13]. While previous studies have fabricated thin films using various deposition techniques such as evaporation [14], chemical methods [14], or electro-chemical processes [16, 17], the thin films in this study were deposited via magnetron sputtering. It is worth noting that prior research has shown that magnetic entropy changes and refrigeration capacity of Ni-Mn-Sn and Ni-Mn-Co-Sn films tend to increase with film thickness [18].

Both theoretical and experimental investigations have underscored the distinct differences in physical properties between thin layers and their bulk counterparts [19–21]. For instance, calculations have revealed a magnetic entropy change of 9.57 J/kg·K in Gd films, occurring at a Curie temperature of 294 K, character-ized by a second-order magnetic phase transition under a 5T magnetic field [22]. Le Barman and Kaur have identified Ni-Mn-Sb-Al thin films as potential candi-dates for micro-length scale magnetic refrigeration applications [23], while Mello et al. have similarly highlighted Tb thin films as promising materials for magne-tocaloric effect devices, particularly for intermediate temperature applications [24]. Temperature-dependent magnetization studies have shown that martensitic transfor-mation temperatures exhibit a monotonic increase with rising Cr content, attributed to changes in valence electron concentration and cell volume [25–27].

In the present work, we have determined critical parameters, including the reduced surface coupling and the reduced transition temperature. Additionally, we have quan-tified the magnetic entropy change and relative cooling power, obtained total and partial magnetic hysteresis cycles, and deduced the magnetic coercive field.

## 7.2   Ising Model and Monte Carlo Simulations

The Hamiltonian described the perovskites ferromagnetic thin films with L mono-layers (see Fig. 7.1) using Ising model includes nearest neighbors interactions and external magnetic field $h$ is given by:

$$H = -R \sum_{<i, j>} S_i S_j - \sum_{<i, k>} S_i S_k - \frac{h}{J_B} \sum_i S_i,$$

with $R = J_S/J_B$ and $t = T/J_B$.

Where the first term represents the nearest neighbor interactions with the ferro-magnetic exchange coupling $J_{ij}$ ($J_S$: surface coupling and $J_B$: bulk coupling), and the second summation which is carried out over all the lattice sites denotes the Zeeman energy term. The spin moment takes $S = \pm 2; \pm 1$ and 0.

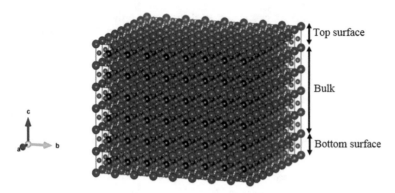

**Fig. 7.1** Schematic representation of a ferromagnetic thin film composed of L distinct monolayers. The bottom and top surfaces have monolayer thickness. Surface spin–spin interaction is denoted by $J_s$. Bulk region thickness is $L - 2$ with exchange coupling $J_b$. Blue, red and black are the A, X and black magnetic ions, respectively

The perovskites ferromagnetic thin films with L monolayers are assumed to reside in the unit cells and the system consists of the total number of spins N = 1600. We have used the Monte Carlo simulation to simulate the Hamiltonian given by Eq. (1). The cyclic free conditions on the lattice of the system following z-axes and boundary condition in (y, x) plane are applied. Monte Carlo update was performed by choosing random spins and then flipped (from current state $S_i$ to opposite state $-S_i$) with Boltzmann based probability with Metropolis algorithm [32].

Our program calculates the following parameters, namely:

The internal energy per site $E$ is of perovskites ferromagnetic thin films, $E = \frac{1}{N} \langle H \rangle$.

The surface and bulk magnetizations of perovskites ferromagnetic thin films are given by:

$$M_S = \left\langle \frac{1}{N_S} \sum_i S_i \right\rangle \text{ and } M_B = \left\langle \frac{1}{N_B} \sum_i S_i \right\rangle \text{ with } N_S = 2xN \text{ and with } N_B = (L-2)xN.$$

The total magnetization of perovskites ferromagnetic thin films is:

$$M = \frac{N_S M_S + N_B M_B}{N_S + N_B}$$

The surface and bulk magnetic susceptibilities of perovskites ferromagnetic thin films are given by:

$$\chi_S = \beta\left(\langle M_S^2 \rangle - \langle M_S \rangle^2\right)$$

$$\chi_B = \beta\left(\langle M_B^2 \rangle - \langle M_B \rangle^2\right)$$

The total susceptibility of perovskites ferromagnetic thin films is:

$$\chi = \beta\left(\langle M^2 \rangle - \langle M \rangle^2\right)$$

where $\beta = \frac{1}{k_B T}$, T denotes the absolute temperature and $k_B$ is the Boltzmann's constant.

The magnetic specific heat of perovskites ferromagnetic thin films is given by:

$$C_m = \frac{\beta^2}{N}\left(\langle E^2 \rangle - \langle E \rangle^2\right)$$

where $\beta = \frac{1}{k_B T}$, T denotes the absolute temperature.

Magnetocaloric effect can be related to the magnetic properties of the material through a thermodynamic Maxwell relation

$$\left(\frac{\partial S}{\partial h}\right)_T = \left(\frac{\partial M}{\partial T}\right)_h$$

The magnetic entropy changes of a material can be calculated from this equation;

$$S(T, h) = \int_0^T \frac{C_m}{T} dT$$

The magnetic entropy changes between $h$ different to zero and $h = 0$ is:

$$\Delta S_m(T, h) = S_m(T, h) - S_m(T, 0) = \int_0^{h_{max}} \left(\frac{\partial M}{\partial T}\right)_{h_i} dh$$

$$= \sum_i \left(\frac{\partial M}{\partial T}\right)_{h_i} \Delta h_i$$

$S_m$ (T,h) and $S_m$ (T,0) are the total entropy in presence and absence of magnetic field, respectively and $h_{max}$ is the maximum applied external magnetic field. $\left(\frac{\partial M}{\partial T}\right)_{h_i}$ is the thermal magnetization for a fixed magnetic field $h_i$.

   The defined parameter of relative cooling power (RCP) described as an area under
the dependence of $\Delta S_m(T)$ on temperature, is a compromise between the magnitude
of the magnetic entropy change and the width of the peak. The expression of relative
cooling power RCP is:

$$RCP = \int_{T_c}^{T_h} \Delta S_m(T)dT$$

where $T_c$ and $T_h$ are the cold and the hot temperatures corresponding to both ends
of the half-maximum value of $\Delta S_m^{max}$, respectively.

## 7.3   Results and Discussion: Surface Effects in Perovskites Ferromagnetic Thin Films

In Fig. 7.2, we have presented the thermal partial and total magnetizations, as well
as the magnetic susceptibilities, for perovskite ferromagnetic thin films with specific
parameters: $L = 6$, $R = +0.1$, and $h = 0.5$. The reduced transition temperatures,
corresponding to surface magnetization ($M_S$), bulk magnetization ($M^B$), and total
magnetization (M), are located at the peaks of the magnetic susceptibilities ($\chi_S$, $\chi_B$,
and $\chi$) and were subsequently determined. The obtained values of $t_C$(surface) $= 4.5$
and $t_C$(bulk) $= 10.55$.
   Figure 7.3, shows, the phase diagram of the system plotted in a ($R = J_S/J_B$, $t_C$
$= T_C/J_B$) plane. Each curve corresponds to a certain film thickness $L = 4, 6, 8$ and
10. We see that the curves intersect at the same abscissa $R_C = 1.6$ and ordinate
point $t_C^B = 12.34$. At those points $t_C$ is independent of the film thicknesses and
becomes similar to 3D infinite bulk system, where the surfaces (001) and (00L) are
not important. According to those results a new definition of the $R_C$ parameter should

**Fig. 7.2** The thermal partial
and total magnetizations and
magnetic susceptibilities of
perovskites ferromagnetic
thin films with $L = 6$,
$R = +0.1$ and $h = 0.5$

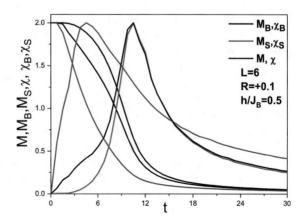

**Fig. 7.3** Phase diagram of
the system plotted in a ($R =$
$J_S/J_B$, $t_C = T_C/J_B$) plane.
Each curve corresponds to a
certain film thickness L = 4,
6, 8 and 10

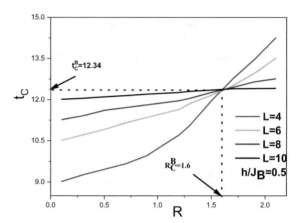

be given: $R_C$ can be defined as the particular R value at which $t_C$ does not depend on
the film thickness and then, $t_C$ is the reduced temperature of the bulk system $t_C^B$. As we
enlarge the value of R the film ordering temperature is raised and when R is greater
than $R_C$ becomes higher than that of the bulk one $t_C^B$. This special point coincides
effectively with that of the semi-infinite system which separates the ordinary from
the extraordinary transitions. For R < $R_C$, $t_C$ is less than the bulk critical temperature
and approaches the last one for large values of L. For R > $R_C$, the system orders at
the surface before it does in the bulk and approaches a limit dependent on the surface
exchange coupling that corresponds to the surface magnetic transition observed in
the corresponding semi-infinite film [34].

Figure 7.4, illustrates, the temperature dependence of magnetic entropy change
$|\Delta S|$ corresponding to different external magnetic field $h/J_B = 0.5, 1.0, 1.5, 2.0$
with film thickness L = 4 (a) and 10(b) for R = + 0.1. The maximum of magnetic
entropy change $|\Delta S^{max}|$, increases with increasing the external magnetic field and
decreases with increasing the thin thickness L (see Fig. 7.4a and b). These results are
comparable with those given by the effective field theory [36]. The reduced transition
temperatures for L = 4 and 10 are $t_C = 9$ and 11.9, respectively.

Temperature dependence of magnetic entropy change $|\Delta S|$ corresponding to
different external magnetic field $h/J_B = 0.5, 1.0, 1.5, 2.0$ with film thickness L = 4
(a) and 10(b) for R = + 2.1 (see Fig. 7.5). The maximum of magnetic entropy change
$|\Delta S^{max}|$, increases with increasing the external magnetic field and decreases also with
increasing the thin thickness L such as given in Fig. 7.5a and b. These results are
comparable with those given by the effective field theory [35]. The reduced transition
temperatures for L = 4 and 10 are $t_C = 14.24$ and 12.45, respectively.

The values of $|\Delta S^{max}|$ with different values of magnetic field for R = + 0.1 and
+ 2.1 are given in Tables 7.1 and 7.2, respectively.

From Tables 7.1 and 7.2, it is evident that the absolute value of the maximum
magnetic entropy change ($|\Delta S^{max}|$) decreases as the reduced surface coupling (R)
increases for L = 4, and it increases as R increases for L = 10.

**Fig. 7.4** Temperature dependence of magnetic entropy change |ΔS| corresponding to different external magnetic field h/J$_B$ = 0.5, 1.0, 1.5, 2.0 with film thickness L = 4 **a** and 10 **b** for R = + 0.1

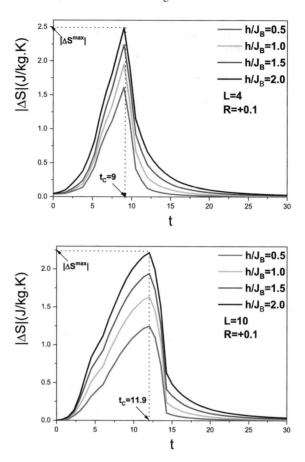

In Fig. 7.6, the temperature dependence of the magnetic entropy change |ΔS| is shown for various film thicknesses (L = 4, 6, 8, 10) with R = + 0.1 (a) and + 2.1 (b) for h/J$_B$ = 1.0. The reduced transition temperature increases with increasing film thickness for R < 1 (Fig. 7.6a) and decreases for R > 1 (Fig. 7.6b). The |ΔS$^{max}$| values decrease with increasing film thickness for both R = + 0.1 and + 2.1.

Figure 7.7 illustrates the external magnetic field dependence of the relative cooling power (RCP) for different film thicknesses (L = 4, 6, 8, 10) with R = + 0.1 (a) and + 2.1 (b). The RCP value increases with increasing magnetic fields and film thickness (Fig. 7.7a) for R = + 0.1, while it decreases with increasing film thickness (Fig. 7.7b) for R = 2.1.

Figure 7.8 displays the external magnetic field dependence of RCP for different values of R = + 0.1 and 2.1, with a fixed film thickness of L = 10. The RCP values increase with increasing values of reduced surface coupling (R), as shown in Fig. 7.8.

Finally, Fig. 7.9 presents the total magnetic hysteresis cycle of the ferromagnetic thin film (a), the magnetic hysteresis cycle of the bulk (b), and the magnetic hysteresis cycle of the surface (c) for different temperatures (t = 1.0, 4.2, 7.2), R = + 0.1, and

**Fig. 7.5** Temperature dependence of magnetic entropy change |ΔS| corresponding to different external magnetic field h/J_B = 0.5, 1.0, 1.5, 2.0 with film thickness L = 4 **a** and 10 **b** for R = + 2.1

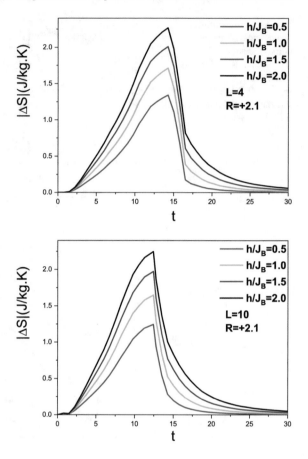

**Table 7.1** The values of $|\Delta S^{max}|$ with diffident values of h/J_B = 0.5, 1, 1.5, 2 for R = + 0.1, L = 4 and 10

| h/J_B | 0.5 | 1 | 1.5 | 2 |
|---|---|---|---|---|
| $|\Delta S^{max}|$ R = + 0.1 L = 4 | 1.61 | 1.96 | 2.24 | 2.5 |
| $|\Delta S^{max}|$ R = + 0.1 L = 10 | 1.25 | 1.62 | 1.94 | 2.22 |

L = 8. The magnetic coercive field decreases with increasing temperature (t = 1.0, 4.2, 7.2) for L = 8. Additionally, the maximum saturation magnetization decreases with increasing temperature.

Figure 7.10, shows the magnetic coercive field of ferromagnetic thin film versus the exchange interactions R with L = 4 and 10. For $R < R_C = 1.6$, the magnetic coercive field increases with increasing the value of film thickness and decreases

**Table 7.2**   The values of $|\Delta S^{max}|$ with diffident values of h/J$_B$ = 0.5, 1, 1.5, 2 for R = + 2.1, L = 4 and 10

| h/J$_B$ | 0.5 | 1 | 1.5 | 2 |
|---|---|---|---|---|
| $\|\Delta S^{max}\|$ R = 2.1 L = 4 | 1.34 | 1.71 | 2.01 | 2.27 |
| $\|\Delta S^{max}\|$ R = 2.1 L = 10 | 1.26 | 1.65 | 1.97 | 2.24 |

**Fig. 7.6**   Temperature dependence of magnetic entropy change $|\Delta S|$ corresponding to different film thickness L = 4, 6, 8, 10 with R = + 0.1 (a) and + 2.1(b) for h/J$_B$ = 1.0

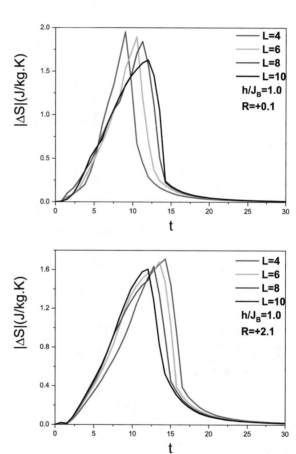

for $R > R_C = 1.6$. $R_C$ is the same obtained in Fig. 7.3 and can be defined as the particular R value at which $t_C$ does not depend on the film thickness and then, $\frac{h_C}{J_B}$ is the reduced magnetic field of the bulk system $\left(\frac{h_C}{J_B}\right)^B$.

**Fig. 7.7** The external magnetic field dependence of RCP corresponding to different film thickness L = 4, 6, 8, 10 with R = + 0.1 (a) and + 2.1(b)

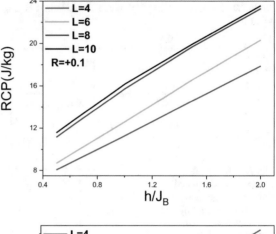

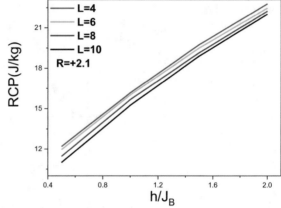

**Fig. 7.8** The external magnetic field dependence of RCP corresponding to different values of R = + 0.1 and 2.1 and film thickness L = 10

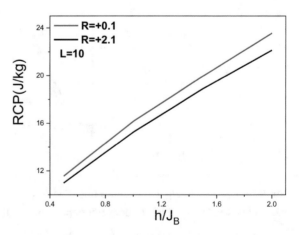

**Fig. 7.9** The total magnetic hysteresis cycle of ferromagnetic thin film **a**, magnetic hysteresis cycle of bulk **b** and magnetic hysteresis cycle of surface **c** for different temperatures t = 1.0, 4.2, 7.2, R = + 0.1 and L = 8

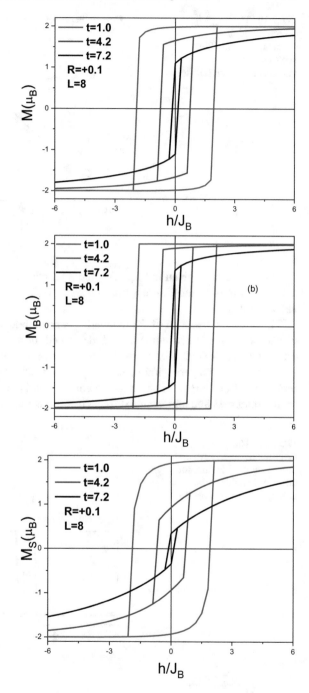

**Fig. 7.10** The magnetic coercive field of ferromagnetic thin film versus the exchange interactions R with L = 4 and 10

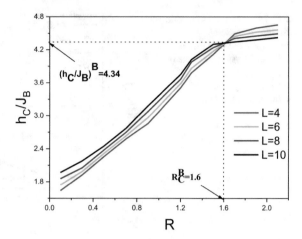

Finally, we have given in Fig. 7.11, the total magnetic hysteresis cycle of ferromagnetic thin film, for different film thickness L = 4, 6, 8 and 10 with R = + 2.1 and t = 4.2.

The observed behavior of the magnetic coercive field decreasing with increasing film thickness for a fixed temperature and reduced surface coupling is consistent with findings in other studies, such as the investigation of superparamagnetic behavior in ultrathin CoNi layers within electrodeposited CoNi/Cu multilayer nanowires, as reported in Ref. [36]. This suggests a common trend in magnetic properties related to film thickness and temperature effects in various magnetic systems.

**Fig. 7.11** The total magnetic hysteresis cycle of ferromagnetic thin film, for different film thickness L = 4, 6, 8 and 10 with R = + 2.1 and t = 4.2

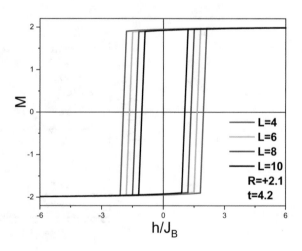

## 7.4 Conclusions

Your study utilizing Monte Carlo simulations has provided valuable insights into the behavior of reduced transition temperature, magnetic entropy change, and relative cooling power in thin ferromagnetic films of varying thickness and surface coupling. The identification of a critical value RC for the R parameter, beyond which tC becomes independent of film thickness, is an important finding.

Furthermore, the observations that $|\Delta S^{max}|$ increases with higher external magnetic fields while decreasing with thinner film thickness are noteworthy. Additionally, the increase in relative cooling power with higher magnetic fields, surface coupling, and film thickness highlights the interplay of these factors in determining the magnetic properties of these thin films. These findings contribute to a better understanding of how various parameters influence the magnetocaloric properties of ferromagnetic thin films.

The magnetic coercive field decreases with increasing the values of temperatures and increases with increasing the value of film thickness for $R < R_C$ and decreases for $R > R_C$. The magnetic coercive field decreases with increasing the film thickness for fixed value of temperature and reduced surface coupling.

## References

1. D. Giratá, A. Hoffmann, O. Arnache, Fe-doping and strain effects on structural and magnetotransport properties in $La_{2/3}Ca_{1/3}Mn_{1-y}Fey O_3$ thin films. Phys. Rev. B **77**, 214430 (2008)
2. S.B. Kansara, D. Dhruv, Z. Joshi, D.D. Pandya, S. Rayaprol, P.S. Solanki, D.G. Kuberkar, N.A. Shah, Structure and microstructure dependent transport and magnetic properties of sol–gel grown nanostructured $La_{0.6}Nd_{0.1}Sr_{0.3}MnO_3$ manganites: role of oxygen, Appl. Surf. Sci. **356**, 1272–1281 (2015)
3. K.A. Gschneidner Jr., V.K. Pecharsky, A.O. Tsokol, Rep. Prog. Phys. **68**, 1479 (2005)
4. A. Smith, C.R.H. Bahl, R. Bjork, K. Engelbrecht, K.K. Nielsen, N. Pryds, Adv. Energy Mater. **2**, 1288 (2012)
5. Y.D. Zhang, P.J. Lampen, T.L. Phan, S.C. Yu, H. Srikanth, M.H. Phan, J. Appl. Phys. **111**, 063918 (2012)
6. M.H. Phana, S.C. Yu, J. Magn. Magn. Mater. **308**, 325–340 (2007)
7. K.G. Sandeman, Scr. Mater. **67**, 566 (2012)
8. V. Franco, J.S. Blazquez, B. Ingale, A. Conde, Annu. Rev. Mater. Res. **42**, 05 (2012)
9. A. Bhattacharyya, S. Giri, S. Majumdar, J. Magn. Magn. Mater. **323**, 1484 (2011)
10. R. Moubah, A. Boutahar, H. Lassri, A. Dinia, S. Colis, B. Hjörvarsson, P.E. Jönsson, Mater. Lett. **175**, 5 (2016)
11. A. Ben Hassine, A. Dhahri, L. Bouazizi, M. Oumezzine, E.K. Hlil, Solid State Commun. **233**, 6 (2016)
12. X. Si, Y. Liu, W. Lei, J. Xu, W. Du, J. Lin, T. Zhou, X. Lu, Solid State Commun. **247**, 27 (2016)
13. A. Portavoce, E. Assaf, C. Alvarez, M. Bertoglio, R. Clérac, K. Hoummada, C. Alfonso, A. Charaï, O. Pilone, K. Hahn, V. Dolocan, S. Bertaina, Appl. Surf. Scie. **437**, 336–346 (2018)
14. T. Haeiwa, M. Matsumoto, IEEE Trans. Magn. **24**, 2055–2059 (1988)
15. J. Wang, Q. Yang, J. Zhou, K. Sun, Z. Zhang, X. Feng, T. Li, Nano Res. **3**, 211–221 (2010)

16. S.K. Zecevic, J.B. Zotovic, S.L. Gojkovic, V. Radmilovic, J. Electroanal. Chem. **448**, 245–252 (1998)
17. P. Decker, H.S. Stein, S. Salomon, F. Brüssing, A. Savan, S. Hamann, A. Ludwig, Thin Solid Films **603**, 262–267 (2016)
18. Rajkumar Modak a, M. Manivel Raja b, A. Srinivasan. J. Magn. Magn. Mater. **448**, 146–152 (2018)
19. D.C. Dunand, P. Müllner, Adv. Mater. **23**, 216–232 (2011)
20. G.A. Malygin, Tech. Phys. **54**, 1782–1785 (2009)
21. V.V. Khovaylo, V.V. Rodionova, S.N. Shevyrtalov, V. Novosad, Phys. Status Solidi **251**, 2104–2113 (2014)
22. K.P. Shinde, B.B. Sinha, S.S. Oh, H.S. Kim, H.S. Ha, S.K. Baik, K.C. Chung, D.S. Kim, S. Jeong, J. Magn. Magn. Mater. **374**, 144–147 (2015)
23. R. Barman, D. Kaur, Vacuum **120**, 22–26 (2015)
24. V.D. Mello, D.H.A.L. Anselmo, M.S. Vasconcelos, N.S. Almeida, Solid State Commun. **268**, 56–60 (2017)
25. Harish Sharma Akkera, Inderdeep Singh, Davinder Kaur. J. Magn. Magn. Mater. **424**, 194–198 (2017)
26. E.K. Abdel-Khalek, A.F. Salem, E.A. Mohamed, J. Alloys. Compnds. **608**, 180–184 (2014)
27. E. Restrepo-Parra, L. Ramos-Rivera, J. Londoño-Navarro, J. Magn. Magn. Mater. **351**, 65–69 (2014)
28. R. Masrour, M. Hamedoun, A. Benyoussef, Appl. Surf. Scie. **258**, 1902–1909 (2012)
29. R. Masrour, M. Hamedoun, K. Bouslykhane a , A. Hourmatallah, N. Benzakour, A. Benyoussef. Appl. Surf. Scie. **255**, 7462–7467 (2009)
30. R. Masrour, M. Hamedoun, A. Benyoussef, Phys. Lett. A **373**, 2071–2074 (2009)
31. M. Hamedoun, K. Bouslykhane, H. Bakrim, A. Hourmatallah, N. Benzakour, R. Masrour, J. Magn. Magn. Mater. **301**, 22–30 (2006)
32. N. Metropolis, A.W. Rosenbluth, M.N. Rosenbluth, A.H. Teller, E. Teller, J. Chem. Phys. **21**, 1087 (1953)
33. M.E.J. Newman, G.T. Barkema, *Monte carlo methods in statistical physics* (Oxf. Univ. Press., Oxford, 1999)
34. J.W. Tucker, E.F. Sarmento, J.C. Cressoni, J. Magn. Magn. Mater. **147**, 24 (1995)
35. Y. Yüksel, Ü. Akıncı, E. Vatansever, Thin Solid Films **646**, 67–74 (2018)
36. X.-T. Tang, G.-C. Wang, M. Shima. J. Appl. Phys. **99**, 123910 (2006)

# Chapter 8
# Effect of Magnetic Field on the Magnetocaloric and Magnetic Properties of Perovskite Orthoferrites

**Abstract** The magnetocaloric effect on $SmFeO_3$ perovskite doped by Mn perovskite were studied by Monte Carlo simulations. The temperature-dependent magnetization is found and shows that the Néel temperature of the weak-ferromagnetic $SmFeO_3$ decreases as Fe ions are substituted by Mn ions. A paramagnetic-to-weak-antiferromagnetic transition with decreasing the temperature is observed and the corresponding Néel temperature essentially decreases as the Mn content increases. The magnetocaloric effect shows two peaks related to magnetic behavior changes, at paramagnetic-like behavior $T_K(K)$ and at Néel temperature $T_N(K)$ of $SmFe_{1-x}Mn_xO_3$. The second phase transition is established. The magnetic entropy changes are given for a several magnetic fields. We have also determined the relative cooling power. Finally, the magnetic hysteresis cycles have been obtained with different dilutions and temperatures values.

## 8.1 Introduction

The magnetoelectric coupling of manganite and perovskite ferrite has been studied by Refs. [1, 2]. On the other hand, manganites are brilliant representatives of materials with giant values of the magnetocaloric effect, which makes them potential candidates for application in magnetic cooling technology [3, 4]. However, the magnetocaloric and magnetic properties of the $SmFe_{0.5}Mn_{0.5}O_3$ complex perovskite have also been studied experimentally [5]. The magnetocaloric effect shows two peaks related to changes in magnetic behavior, at 18 K and at 234 K [5] therefore the thermodynamic properties of $SmMnO_3$ perovskite manganites have been studied [6]. transition, as a function of composition ratio and atomic radius [7, 8].

Wide adiabatic temperature range, magnetic entropy change and high relative cooling power are the key parameters to distinguish a magnetic material to be used in magnetic refrigeration technology [9, 10]. Another advantage of manganite-based materials is that the control temperature is also generally higher, which is also beneficial for the technological prospects [11, 12]. The magnetocaloric properties then at equilibrium of Ising systems have been studied using effective field theory [13].

© The Author(s), under exclusive license to Springer Nature Switzerland AG 2024
R. Masrour, *Magnetoelectronic, Optical, and Thermoelectric Properties of Perovskite Materials*, SpringerBriefs in Materials,
https://doi.org/10.1007/978-3-031-48967-9_8

These classes of materials are very topical insofar as doped analogues have already "exploded" into the limelight in the new field of spintronics [9, 10]. The magnetocaloric effect can be defined as the thermal response of magnetic materials due to changes in the applied magnetic field [11–13]. The crystal structure and magnetic properties of rare earth perovskite single crystal $SmFe_{0.5}Mn_{0.5}O_3$ have been studied [14]. DC magnetization measurements of $SmFeO_3$ were performed using a physical property measurement system (Quantum Design, PPMS-9), zero field cooling and field cooling processes were used to acquire the temperature dependence of the magnetization [15]. A class of rare earth orthoferrites $RFeO_3$ (R is a rare earth element) exhibits striking physical properties of spin switching and magnetization reversal induced by temperature and/or applied magnetic field [16].

In the present work, we study the magnetic properties of the $SmFe_{1-x}Mn_xO_3$ complex perovskite using Monte Carlo simulations, and its behavior in terms of magnetocaloric effect at various magnetic field variation intensities and x-dilutions. The relative cooling power is deduced for different dilutions x and for different external magnetic fields. Finally, the magnetic coercive field is calculated for different dilution values x and different temperatures.

## 8.2   Ising Model and Monte Carlo Study

We have used the a $= 5.3615$ Å, b $= 5.6253$ Å and c $= 7.5189$ Å lattice parameters [5] of $SmFe_{0.5}Mn_{0.5}O_3$ perovskite to plot the Fig. 8.1, and it is assumed to reside in the unit cells and the system consists of the total number of spins $N = N_{Sm}{}^{3+} + N_{Fe}{}^{3+} + N_{Mn}{}^{3+}$, with $N_{Sm}{}^{3+}$, $N_{Fe}{}^{3+}$ and $N_{Mn}{}^{3+}$ are the number sites of $Sm^{3+}$, $Fe^{3+}$ and $Mn^{3+}$, respectively such given in Fig. 8.1. The Hamiltonian of the $SmFe_{0.5}Mn_{0.5}O_3$ with a ferrimagnetic spin-5/2, 5/2 and 2 configuration of $Sm^{3+}$, $Fe^{3+}$ and $Mn^{3+}$, respectively includes nearest neighbors' interactions and external magnetic field H is given as:

$$H' = - J_{Sm-Sm} \sum_{<i,j>} S_i S_j - J_{Sm-Fe} \sum_{<i,k>} S_i \sigma_k - J_{Sm-Mn}$$

$$\sum_{<i,n>} S_i q_n - J_{Fe-Fe} \sum_{<k,m>} \sigma_k \sigma_m - J_{Mn-Fe}$$

$$\sum_{<k,n>} \sigma_k q_n - J_{Mn-Mn} \sum_{<n,l>} q_n q_l - H \left( \sum_i S_i + \sum_k \sigma_k + \sum_n q_n \right)$$

The expression of exchange interaction $J_{Mn-Fe}$ is given according to $J_{Fe-Fe}$, $J_{Mn-Mn}$ and $x$ by:

$$J_{Mn-Fe} = \frac{1}{4} \left[ \sum_{m=0}^{2} p_n(m) (J_i J_j)^{m/2} \right]$$

**Fig. 8.1** $SmFe_{1-x}Mn_xO_3$ perovskite

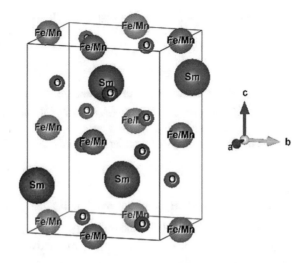

In previous works [17, 18], the probability $p_n(m) = \frac{n!}{m!(2-m)}x^m(1-x)^{n-m}$ that an ion of coordination n = 2 be linked with $m$ ions, each of probability (1-x).

$J_i = J_j = J_{Fe}$ pour m = 0,

$J_i = J_{Fe}$ and $J_j = J_{Mn}$ pour m = 1,

$J_i = J_j = J_{Mn}$ pour m = 2.

We have selected the absolute values of exchange interactions $J_{Sm-Sm} = + 11.4$ K, $J_{Sm-Fe} = + 53.0$ K , $J_{Sm-Mn} = + 1.0$ K, $J_{Fe-Fe} = + 65.0$ K and $J_{Mn-Mn} = + 15.0$ K used in the present work obtained by Holland and Brown [19]. The obtained values of $N_{Sm}^{3+}$, $N_{Fe}^{3+}$ and $N_{Mn}^{3+}$ ions in $SmFe_{1-x}Mn_xO_3$ are given in Table 8.1.

The Monte Carlo simulations update were performed by choosing random spins and then flipped (from current state $S_i$ (or $\sigma_i$ or $q_i$) to opposite state $-S_i$(or $-\sigma_i$ or $-q_i$) with Boltzmann based probability. In general, this can be done using the conventional Metropolis algorithm [20] i.e. $P_{metro} = \exp(-\Delta E/k_B T)$ where $\Delta E$ is the energy difference between the before and the after flip and where T denotes the absolute temperature and $k_B$ is the Boltzmann's constant. However, since the actual Metropolis suffers from large correlation time especially close to the critical point [21]. We have used the cyclic boundary conditions on the lattice were imposed and the configurations were generated by sequentially traversing the lattice and making

**Table 8.1** Numbers of $Sm^{3+}$, $Fe^{3+}$ and $Mn^{3+}$ions in $SmFe_{1-x}Mn_xO_3$ used in our calculations for a several values of x

| Ions/x | $Sm^{3+}$ | $Fe^{3+}$ | $Mn^{3+}$ |
|--------|-----------|-----------|-----------|
| 0 | 150 | 168 | 0 |
| 0.1 | 150 | 148 | 20 |
| 0.5 | 150 | 88 | 80 |
| 0.9 | 150 | 18 | 150 |
| 1 | 150 | 0 | 168 |

single-spin flip attempts. The number of configurations used in this chapter is $10^5$ configurations (with the time interval between configurations at least 1 Monte Carlo step). The magnetization per spin of each ions contain the $Sm^{3+}$, $Fe^{3+}$ and $Mn^{3+}$ is

$$M = \left\langle \frac{1}{N_{Sm^{3+}}} \sum_i S_i \right\rangle + \left\langle \frac{1}{N_{Fe^{3+}}} \sum_i \sigma_i \right\rangle + \left\langle \frac{1}{N_{Mn^{3+}}} \sum_i q_i \right\rangle.$$

The magnetic contribution to the specific heat of $SmFe_{1-x}Mn_xO_3$ is given by: $C_m = \frac{\beta^2}{N}\left(\langle E^2 \rangle - \langle E \rangle^2\right)$ with $E = \frac{1}{N}\langle H' \rangle$ is the internal energy per site and $\beta = \frac{1}{k_BT}$, with T denotes the absolute temperature.

The magnetic entropy of a material can be calculated from this equation.

$$S_m(T, H) = \int_0^T \frac{C_{m,H}}{T'} dT' \quad \text{and the magnetic entropy change is } \Delta S_m(T, H) =$$

$\int_0^{H_{max}} \left(\frac{\partial M}{\partial T}\right)_{H_i} dH$, with $H_{max}$ is the maximum applied field. $\left(\frac{\partial M}{\partial T}\right)_{H_i}$ is the thermal magnetization for a fixed magnetic field $H_i$. The adiabatic temperature is given by $\Delta T_{ad} = -T \frac{\Delta S_m}{C_{p,H}}$. The defined parameter of relative cooling power described as an area under the dependence of $\Delta S_m(T)$ on temperature, is a compromise between the magnitude of the magnetic entropy change and the width of the peak. The expression of relative cooling power RCP is $RCP = \int_{T_c}^{T_h} \Delta S_m(T) dT$, where $T_N$ and $T_h$ are the cold and the hot temperatures corresponding to both ends of the half-maximum value of $\Delta S_m^{max}$, respectively.

## 8.3 Results and Discussion: Magnetocaloric Effect and Magnetic Properties of Perovskite Orthoferrites

The magnetocaloric effect in $SmFe_{1-x}Mn_xO_3$ complex perovskite have been investigated using the Monte Carlo simulation. We have presented in Fig. 8.2a–c, the thermal magnetization for $SmMnO_3$, $SmFe_{0.5}Mn_{0.5}O_3$ and $SmFeO_3$, respectively with H = 6 T. From magnetization versus temperature, we observe two magnetic orderings, the first one at 40 K for x = 0.5 and 84 K for x = 0 due to $Sm^{3+}$, and the other one at $T_N = 240$ K for x = 0.5 and 400 K for x = 0 is the anti-ferromagnetic long-range ordering. The critical temperature $T_K$ and the transition temperatures are obtained for different dilutions x of $SmFe_{1-x}Mn_xO_3$ is given in Table 8.2.

The obtained results are near to those obtained in previous works [5, 14], this difference may be explained by the different magnetic atoms take in the elementary cell and the external magnetic field applied for our system. The critical temperature $T_K$ corresponding to this weak-antiferromagnetic transition principally decreases as Fe ions are substituted by Mn ions. These results are comparable with those given by experiment results [22].

Figure 8.3a–c, shows the Mn content dependence of $T_N$ and $T_K$, respectively of $SmFe_{1-x}Mn_xO_3$ with H = 6 T. The Fig. 8.3a, b, show that as the Mn content increases both $T_N$ and $T_K$ decreases, then may be explained by the distortion taken place as a

**Fig. 8.2** The thermal magnetization for SmMnO$_3$ (**a**), SmFe$_{0.5}$Mn$_{0.5}$O$_3$ (**b**) and SmFeO$_3$ (**c**) with J$_{Sm-Sm}$ = +11.4 K, J$_{Sm-Fe}$ = +53.0 K, J$_{Sm-Mn}$ = +1.0 K, J$_{Fe-Fe}$ = +65.0 K, J$_{Mn-Mn}$ = +15.0K and H = 6.0 T

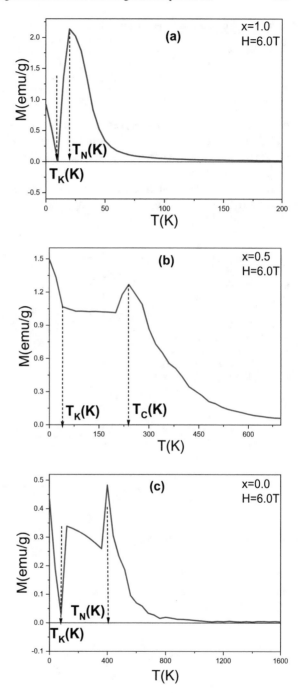

**Table 8.2** The values of $T_K$ and $T_N$ in $SmFe_{1-x}Mn_xO_3$ for x = 0, 0.5 and 1

| x | 0 | | 0.5 | | 1 | |
|---|---|---|---|---|---|---|
| $T_K(K)$ and $T_N(K)$ | $T_K$ | $T_N$ | $T_K$ | $T_N$ | $T_K$ | $T_N$ |
| | 84 | 400 | 40 | 240 | 9.3 | 21 |

result of the increased Mn ions present in the $SmFe_{1-x}Mn_xO_3$ and the decrease of $T_K$ and $T_N$ is due to the interaction between $Fe^{3+}$ and $Mn^{3+}$. The transition from weak ferromagnetic to paramagnetic is also observed.

The decrease may be attributed to the distortion taken place as a result of the increased Mn ions present in the sample. The effect of the substitution of Fe ions by Mn ions in $SmFeO_3$ on the double exchange and the critical temperatures is analyzed and discussed [22].

**Fig. 8.3** Mn content dependence of Néel temperature $T_N(K)$ (**a**) and at paramagnetic-like behaviour $T_K(K)$ (**b**) of $SmFe_{1-x}Mn_xO_3$ for $J_{Sm-Sm} = +11.4$ K, $J_{Sm-Fe} = +53.0$ K, $J_{Sm-Mn} = +1.0$ K, $J_{Fe-Fe} = +65.0$ K, $J_{Mn-Mn} = +15.0$ K and H = 6.0 T

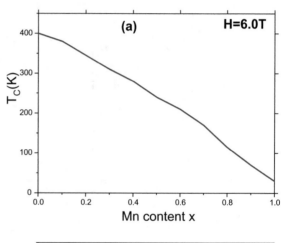

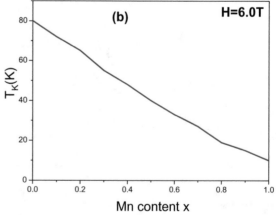

Figure 8.4, illustrates, the temperature dependence of the magnetic entropy changes for $SmFe_{0.5}Mn_{0.5}O_3$ with different external magnetic field H = 6, 12, 18, 24 and 30 T. The expected behavior of change of entropy is observed, which is the increased reduction in entropy with increasing applied field. The maximum of magnetic entropy change around $T_N$ was $3 \, J.\,kg^{-1}K^{-1}$ at the temperature of T = 240 K under the high external magnetic field of 30 T. At low temperatures, we also observed a significant change in entropy, in a variation of the applied field 30 T. The increased reduction in entropy with increasing applied field is also observed. The obtained results are comparable with those given in previous work [23]. The maximum of magnetic entropy change $-\Delta S^{max}$ increases with increasing the external magnetic field with x = 0.5 such as given in Fig. 8.5. This behavior is already explained by Ref. [5], because the upper magnetic fields easily align the magnetic moments, thus reducing the entropy. The same behavior is observed in experimental study of magnetocaloric and magnetic properties of $SmFe_{0.5}Mn_{0.5}O_3$ perovskite [5]. The field dependence of relative cooling power (RCP) for $SmFe_{0.5}Mn_{0.5}O_3$ with different temperatures is given in Fig. 8.6. The value of RCP increases with increasing the temperatures values. We have given in Fig. 8.7a, b, the magnetic hysteresis cycles for (x = 0.1, 0.5, 0.9, T = 100 K) and (T = 100, 200 300 K, x = 0.5), respectively. The magnetic coercive field increases with increasing the dilution x such as given in Fig. 8.7a and decreases with increasing the temperatures values (see Fig. 8.7b). The magnetic coercive field decreases with increasing the dilutions x and temperatures values. This result for dilution x is comparable with that given in Ref. [24] and for temperatures effect is comparable with that given in Refs. [14, 25]. The system is the antiferromagnetic for the temperature below $T_N$ and becomes paramagnetic for higher temperatures (T ≥ $T_N$).

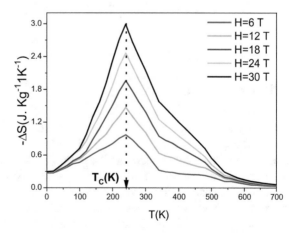

**Fig. 8.4** The temperature dependence of the magnetic entropy changes for $SmFe_{0.5}Mn_{0.5}O_3$ with different external magnetic field H = 6, 12, 18, 24, 30 T with $J_{Sm-Sm} = +11.4$ K, $J_{Sm-Fe} = +53.0$ K, $J_{Sm-Mn} = +1.0$ K, $J_{Fe-Fe} = +65.0$ K and $J_{Mn-Mn} = +15.0$ K

**Fig. 8.5** The external magnetic field dependence of the maximum magnetic entropy change for SmFe$_{0.5}$Mn$_{0.5}$O$_3$ with J$_{Sm-Sm}$ = +11.4 K, J$_{Sm-Fe}$ = +53.0 K, J$_{Sm-Mn}$ = +1.0 K, J$_{Fe-Fe}$ = +65.0 K, J$_{Mn-Mn}$ = +15.0 K, and T = 240 K

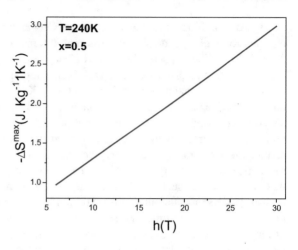

**Fig. 8.6** The field dependence of relative cooling power (RCP) for SmFe$_{1-x}$Mn$_x$O$_3$ with J$_{Sm-Sm}$ = +11.4 K, J$_{Sm-Fe}$ = +53.0 K, J$_{Sm-Mn}$ = +1.0 K, J$_{Fe-Fe}$ = +65.0 K, J$_{Mn-Mn}$ = +15.0 K and x = 0.5

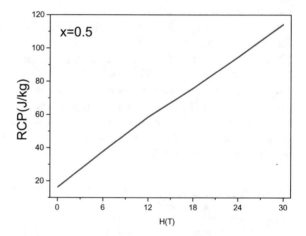

## 8.4   Conclusions

The magnetic properties and the magnetocaloric effect in the complex SmFe$_{1-x}$Mn$_x$O$_3$ perovskite are studied using Monte Carlo study. The critical temperature T$_K$ and the Néel temperature T$_N$ in SmFe$_{1-x}$Mn$_x$O$_3$ with different dilutions x has been found. The both temperatures T$_K$ and T$_N$ are decrease with increasing of dilution x. The paramagnetic weak-antiferromagnetic Néel temperature and the spin reorientation from weak-paramagnetic-to-antiferromagnetic Néel temperature both decrease systematically with increasing Mn content. The increasing of dilution $x$ decreases the maximum of magnetic entropy. The large magnetocaloric effect, relatively high RCP, high magnetization, and low cost jointly make the present compound a promising candidate for magnetic refrigeration near room temperature.

**Fig. 8.7** The magnetic hysteresis cycles for $x = 0.1$, 0.5, 0.9, T = 100 K (**a**) and T = 100, 200 300 K, $x = 0.5$ (**b**) for $SmFe_{1-x}Mn_xO_3$ with $J_{Sm-Sm} = +11.4$ K, $J_{Sm-Fe} = +53.0$ K, $J_{Sm-Mn} = +1.0$ K, $J_{Fe-Fe} = +65.0$ K and $J_{Mn-Mn} = +15.0$ K

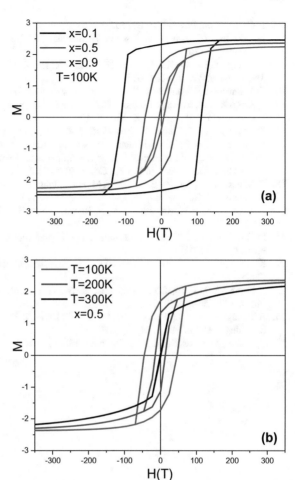

## References

1. P. Tolédano, Phys. Rev. B **79**, 094416 (2009)
2. P. Mandal, V.S. Bhadram, Y. Sundarayya, C. Narayana, A. Sundaresan, C.N. Rao, Phys. Rev. Lett. **107**, 137202 (2011)
3. A.M. Aliev, A.G. Gamzatov, K.I. Kamilov, A.R. Kaul, N.A. Babushkina, Appl. Phys. Lett. **101**, 172401 (2012)
4. C. Martin, A. Maignan, M. Hervieu, B. Raveau, Phys. Rev. B **60**, 12191 (1999)
5. M.C. Silva-Santana, C.A. daSilva, P. Barrozo, E.J.R. Plaza, L. de los Santos Valladares, N.O. Moreno. J. Magn. Magn. Mater. **401**, 612–617 (2016)
6. R. Choithrani, N.K. Gaur, R.K. Singh. J. Phys.:Condens. Matter **20**, 415201 (2008)
7. A. Rebello, R. Mahendiran, Appl. Phys. Lett. **93**, 232501 (2008)
8. J. Mira, J. Rivas, F. Rivadulla, C. Vázquez-Vázquez, M.A. López-Quintela, Phys. Rev. B **60**, 2998 (1999)
9. I. Hussain, M.S. Anwar , S.N. Khan, J.W. Kim, K.C. Chung, B.H. Koo. J. Alloy. Compd. **694**, 815e822 (2017)

10. L.-W. Li, Chin. Phys. B **25**, 037502 (2016)
11. I. Hussain, M.S. Anwar, S.N. Khan, A. Shahee, Z.U. Rehman, B.H. Koo. Inter. Ceram **43**, 10080–10088 (2017)
12. C. Wang, Y. Hu, Q. Hu, F. Chen, M. Zhang, X. He, Z. Li, D. Wang, Q. Cao, Y. Du. Magn. Magn. Mater. **439**, 13–16 (2017)
13. E. Vatansever, Ü. Akinci, Y. Yüksel, Phys. A **479**, 563–571 (2017)
14. J. Kang, X. Cui, Y. Fang, J. Zhang, Solid State Commun. **248**, 101 (2016)
15. J. Kang, Y. Yang, X. Qian, X. Kai, X. Cui, Y. Fang, V. Chandragiri, B. Kang, B. Chen, A. Stroppa, S. Cao, J. Zhang, W. Ren, IUCr. J. **4**, 598–603 (2017)
16. S. Cao, H. Zhao, B. Kang, J. Zhang, W. Ren, Sci. Rep. **4**, 5960 (2014)
17. T.C. Gibb, Magnetic exchange interactions in perovskite solid solutions. Part 1. Iron-57 and $^{151}$Eu Mossbauer spectra of $EuFe_{1-x}Co_xO_3$ (0<x<1). J. Chem. Soc. Dalton. Trans. 873 (1983)
18. R. Masrour, A. Jabar, A. Benyoussef, M. Hamedoun, E.K. Hlil, Indian J. Phys. **90**, 819 (2016)
19. W.E. Holland, H.A. Broun, Phys.-Stat. Sol. (a) **10**, 249 (1972)
20. N. Metropolis, A.W. Rosenbluth, M.N. Rosenbluth, A.H. Teller, E. Teller, J. Chem. Phys. **21**, 1087 (1953)
21. M.E.J. Newman, G.T. Barkema, *Monte Carlo Methods in Statistical Physics* (Oxford University Press, Oxford, 1999)
22. K. Bouziane, A. Yousif, I.A. Abdel-Latif, K. Hricovini, C. Richter, J. Appl. Phys. **97**, 10A504 (2005)
23. Y. Nagata, S. Yashiro, T. Mitsuhashi, A. Koriyama, H. Kawashima, H. Samata, J. Magn. Magn. Mater. **237**, 250–260 (2001)
24. A.T. Raghavender, D. Pajic, K. Zadro, T. Milekovic, P. Venkateshwar Rao, K.M. Jadhav, D. Ravinder. J. Magn. Magn. Mater. **316**(2007), 1–7
25. E. Kantar, M. Ertaş, Solid State Commun. **188**, 71 (2014)

# General Conclusion

We have use the DFT and Monte Carlo simulations to study the magnetic, electronic and magnetocaloric properties of $La_{0.75}Sr_{0.25}MnO_3$ and of $Pr_{0.65}Sr_{0.35}MnO_3$ perovskites. The electronic and magnetic properties were studied using the DFT with GGA, GGA + U and TB-mBJ. We have used the Monte Carlo method to determine the magnetocaloric properties of both compounds. The spin polarization for two compounds shows a half metallic character with the ferromagnetic coupling of the Mn spin. The magnetic properties of these compounds exhibit a single transition from the FM state to the PM state. The values of critical temperature were found. These results are comparable to those obtained experimentally. The results obtained from magnetic entropy and specific heat are processed and analyzed. The maximum magnetic entropy is at the Curie temperature. Its increases with increasing magnetic field. The adiabatic temperature and the specific heat are treated. The variation of the RCP values as a function of the values of the magnetic fields and the temperature is found. RCP values increase with increasing magnetic field. But in magnetic materials we cannot locate the f or d bands with this approach. On the other hand, the GGA + U approximation the exact choice for the value of U allows a localization of the bands (d or f), a total moment, and a fixed load. The main objective of this part was the theoretical study of the Sr effect on electronic, magnetic and magnetocaloric properties in $BaFeO_3$ compound. Analysis of total and partial DOS and the value of difference energy showed that the system $Ba_{1-x}Sr_xFeO_3$ has a ferromagnetic behavior and half metallic character. The Sr substituted $BaFeO_3$ leads to decrease the magnetic moment and Curie temperature. Moreover, we used the Monte Carlo simulation to study the magnetic properties such as magnetization, magnetic entropy, susceptibility and specific heat. The maximum of these magnetic properties is decrease with Sr substitution. The $Ba_{0.8}Sr_{0.2}FeO_3$ shows the second-order PM-FM phase transition around 53 K with a large magnetic entropy $4\ J.K^{-1}.kg^{-1}$ for $x = 0.2$ at $H = 5T$. We have studied the electronic, magnetic, thermoelectric and magnetocaloric properties of perovskite $GdCrO_3$. The band structure shows that the compound $GdCrO_3$ is a semiconductor with a direct gap. For the optical properties we have made qualitative

R. Masrour, *Magnetoelectronic, Optical, and Thermoelectric Properties
of Perovskite Materials*, SpringerBriefs in Materials,
https://doi.org/10.1007/978-3-031-48967-9

studies for certain optical constants, i.e., the dielectric function, the absorption coefficient and the spectrum of reflectivity. The thermoelectric response is evaluated by calculating electrical conductivity, thermal conductivity and Seebeck coefficient. The Néel and spin reorientation temperatures of $GdCrO_3$ compound were found using Monte Carlo study. The maximum of magnetic entropy increases with increasing the external magnetic field. The values of $h_C$ and $M_r$ decrease with increasing the temperature. The electronic structure, optical and photovoltaic properties are investigated and analyzed. The quantum efficiency, the short-circuit current and Open Circuit Voltage revealed by recent experimental studies have been confirmed by our theoretical investigations. Monte Carlo simulations have allowed us to understand and obtain the behavior of ferroelectric properties. $BiFeO_3$ exhibits a second order phase transition ferroelectric to paraelectric in the vicinity of the Curie temperature $T_c = 1103$ K. As the annealing temperature rises, the $E_C$ and $P_r$ gradually decreases. The values of Ec and Pr increase with increasing the value of $N_{Layers}$ until they are saturated. Finally, the density functional theory based on FPLAPW calculation is carried out to investigate the photocatalytic electronic and optical properties of Fe doped in $CsBrO_3$ compound. The calculation of formation energy showed that $CsBrO_3$ has great structural stability. The $CsBrO_3$ is a semiconductor with a wide direct gap 4.24 eV. The dielectric function, absorption, reflectivity and optical conductivity are calculated. The highest peaks in the imaginary part of dielectric function are observed due to the O-p, Br-p and Br-d state in valence band and Br-s, Br-p and O-p state in conduction band in $CsBrO_3$. The band edges alignment indicates that both conduction band minimum and valence band maximum achieve the requirement of a photo-catalyst for water splitting.